AF792603

THEORY OF NONLINEAR ACOUSTICS IN FLUIDS

FLUID MECHANICS AND ITS APPLICATIONS
Volume 67

Series Editor: R. MOREAU
MADYLAM
Ecole Nationale Supérieure d'Hydraulique de Grenoble
Boîte Postale 95
38402 Saint Martin d'Hères Cedex, France

Aims and Scope of the Series

The purpose of this series is to focus on subjects in which fluid mechanics plays a fundamental role.

As well as the more traditional applications of aeronautics, hydraulics, heat and mass transfer etc., books will be published dealing with topics which are currently in a state of rapid development, such as turbulence, suspensions and multiphase fluids, super and hypersonic flows and numerical modelling techniques.

It is a widely held view that it is the interdisciplinary subjects that will receive intense scientific attention, bringing them to the forefront of technological advancement. Fluids have the ability to transport matter and its properties as well as transmit force, therefore fluid mechanics is a subject that is particulary open to cross fertilisation with other sciences and disciplines of engineering. The subject of fluid mechanics will be highly relevant in domains such as chemical, metallurgical, biological and ecological engineering. This series is particularly open to such new multidisciplinary domains.

The median level of presentation is the first year graduate student. Some texts are monographs defining the current state of a field; others are accessible to final year undergraduates; but essentially the emphasis is on readability and clarity.

Theory of Nonlinear Acoustics in Fluids

by

BENGT O. ENFLO

Kungl. Tekniska Högskolan,
Department of Mechanics, Stockholm, Sweden

and

CLAES M. HEDBERG

Blekinge Tekniska Högskola,
Department of Mechanical Engineering, Karlskrona, Sweden

KLUWER ACADEMIC PUBLISHERS
DORDRECHT / BOSTON / LONDON

A C.I.P. Catalogue record for this book is available from the Library of Congress.

ISBN 1-4020-0572-5

Published by Kluwer Academic Publishers,
P.O. Box 17, 3300 AA Dordrecht, The Netherlands.

Sold and distributed in North, Central and South America
by Kluwer Academic Publishers,
101 Philip Drive, Norwell, MA 02061, U.S.A.

In all other countries, sold and distributed
by Kluwer Academic Publishers,
P.O. Box 322, 3300 AH Dordrecht, The Netherlands.

Printed on acid-free paper

All Rights Reserved
© 2002 Kluwer Academic Publishers
No part of this work may be reproduced, stored in a retrieval system, or transmitted
in any form or by any means, electronic, mechanical, photocopying, microfilming, recording
or otherwise, without written permission from the Publisher, with the exception
of any material supplied specifically for the purpose of being entered
and executed on a computer system, for exclusive use by the purchaser of the work.

Printed in the Netherlands.

Contents

Preface **xi**

1 Introduction **1**
1.1 The place of acoustics in fluid mechanics 1
1.2 Nonlinear acoustics before 1950 2
1.3 Special phenomena in nonlinear acoustics 4
1.3.1 Common theoretical description of nonlinear acoustics phenomena 4
1.3.2 Generation and propagation of higher harmonics in travelling waves 5
1.3.3 Generation and propagation of combination frequency travelling waves 7
1.3.4 Propagation of travelling short pulses and N-waves 8
1.3.5 Propagation of limited sound beams 9
1.3.6 Waves in closed tubes 9

2 Physical theory of nonlinear acoustics **11**
2.1 Basic theory of motion of a diffusive medium 12
2.1.1 Conservation of mass; the continuity equation. . 13
2.1.2 Conservation of momentum. Navier-Stokes equations. 14
2.1.3 Conservation of energy. 15
2.1.4 Ideal fluid equation of state 18
2.2 Derivation of the three dimensional wave equation of nonlinear acoustics (Kuznetsov's equation) 20
2.3 Wave equations of nonlinear acoustics 24

2.3.1 Burgers' equation 24
2.3.2 Generalized Burgers' equation 27
2.3.3 The KZK equation 28

3 Basic methods of nonlinear acoustics 31
3.1 Solution methods to the Riemann wave equation 31
3.1.1 Physical interpretation of the Riemann equation 31
3.1.2 Continuous wave solution 33
3.1.3 Shock wave solution 35
3.1.4 Rule of equal areas 38
3.1.5 Prediction of wave behaviour from area differences 42
3.2 Exact solution of Burgers' equation 45
3.2.1 The Cole-Hopf solution of Burgers' equation . . 46
3.2.2 Burgers' equation with vanishing diffusivity . . 49

4 Nonlinear waves with zero and vanishing diffusion 53
4.1 Short pulses 53
4.1.1 Triangular pulses 53
4.1.2 N-waves 56
4.2 Sinusoidal waves 59
4.2.1 Continuous solution 59
4.2.2 The Bessel-Fubini solution 61
4.2.3 Sawtooth solution 62
4.2.4 The one saddle-point method 66
4.2.5 Time reversal 72
4.3 Modulated Riemann waves 76
4.3.1 Direct method for bifrequency boundary condition 76
4.3.2 The one saddle-point method for bifrequency boundary condition 79
4.3.3 Characteristic multifrequency waves 84

5 Nonlinear plane diffusive waves 93
5.1 Planar N-waves 93
5.1.1 Shock solution 93
5.1.2 Old-age solution 98

5.1.3 The old-age solution found by an alternative method 100
5.2 Planar harmonic waves. The Fay solution. 105
5.2.1 Derivation of Fay's solution from the Cole-Hopf solution 105
5.2.2 Direct derivation of Fay's solution 109
5.2.3 Proof that Fay's solution satisfies Burgers' equation 110
5.2.4 Some notes on Fay's solution 112
5.3 Planar harmonic waves. The Khokhlov-Soluyan solution 114
5.3.1 Derivation of the Khokhlov-Soluyan solution . . 114
5.3.2 Comparison between the Fay and the Khokhlov-Soluyan solutions 118
5.3.3 Comparison between the Khokhlov-Soluyan solution and the sawtooth solution 122
5.4 Planar harmonic waves. The exact solution. 125
5.4.1 Recursion formulae for the Fourier series of the exact solution 125
5.4.2 Solving recursion formulae by discrete integration 129
5.4.3 Comparison of Fourier coefficients in the Bessel-Fubini solution, the Fay solution and the exact solution 133
5.5 Multifrequency waves 137
5.5.1 Expressions for multifrequency solutions 137
5.5.2 Bifrequency solutions and creation of combination frequencies 141

6 Nonlinear cylindrical and spherical diffusive waves 149
6.1 Dimensionless generalized Burgers' equations 150
6.2 Cylindrical N-waves 153
6.2.1 Evolution of an initial cylindrical N-wave 153
6.2.2 Four-step procedure for finding the asymptotic solution 154
6.3 The decay of a shockwave from a supersonic projectile 166
6.3.1 Linear theory of the wave from a supersonic projectile 167

6.3.2 Nonlinear theory of the wave from a supersonic projectile 173
6.4 Periodic cylindrical and spherical waves 186
6.4.1 Spherical periodic waves 187
6.4.2 Cylindrical periodic waves 193

7 Nonlinear bounded sound beams **199**
7.1 The KZK equation 201
7.1.1 Dimensionless KZK equation 201
7.1.2 Transformation of the KZK equation to a generalized Burgers' equation 205
7.1.3 Expansion of the solution around the center of the beam 205
7.1.4 Solution for a circular beam 208
7.2 Propagation of a shock wave in a sound beam 210
7.2.1 Determination of the boundary condition from the series solution 210
7.2.2 Solution of generalized Burgers' equation 214
7.2.3 Conditions for shock preservation 216

8 Nonlinear standing waves in closed tubes **219**
8.1 Nonlinear and dissipative effects at non-resonant and resonant driving frequencies 221
8.1.1 Linear theory of standing waves 222
8.1.2 Discussion of the small numbers in the problem of nonlinear standing waves 224
8.2 Equations of nonlinear standing waves 226
8.2.1 Perturbation solution and boundary conditions of Kuznetsov's equation 226
8.2.2 Equations of resonant standing waves 230
8.3 Steady-state resonant vibrations in a non-dissipative medium 231
8.3.1 Continuous solution 232
8.3.2 Shock solution 234
8.3.3 The Q-factor 237
8.4 Steady-state resonant vibrations in a dissipative medium 238

8.4.1 Mathieu equation solution 238
8.4.2 Perturbation theory. Matching outer and inner solutions 240
8.4.3 Perturbation theory. Uniform solution 244
8.5 An example of velocity field in a resonator 247

Bibliography **251**

Name index **271**

Subject index **279**

Preface

The aim of the present book is to present theoretical nonlinear acoustics with equal stress on physical and mathematical foundations. We have attempted explicit and detailed accounting for the physical phenomena treated in the book, as well as their modelling, and the formulation and solution of the mathematical models. The nonlinear acoustic phenomena described in the book are chosen to give physically interesting illustrations of the mathematical theory. As active researchers in the mathematical theory of nonlinear acoustics we have found that there is a need for a coherent account of this theory from a unified point of view, covering both the phenomena studied and mathematical techniques developed in the last few decades.

The most ambitious existing book on the subject of theoretical nonlinear acoustics is "Theoretical Foundations of Nonlinear Acoustics" by O.V. Rudenko and S.I. Soluyan (Plenum, New York, 1977). This book contains a variety of applications mainly described by Burgers' equation or its generalizations. Still adhering to the subject described in the title of the book of Rudenko and Soluyan, we attempt to include applications and techniques developed after the appearance of, or not included in, this book. Examples of such applications are resonators, shockwaves from supersonic projectiles and travelling of multifrequency waves. Examples of such techniques are derivation of exact solutions of Burgers' equation, travelling wave solutions of Burgers' equation in non-planar geometries and analytical techniques for the nonlinear acoustic beam (KZK) equation. The analytical techniques are developed from first principles, i.e. from the continuity equation, Navier-Stokes' equations, the heat conduction equation and the constitutive equation of the medium, in which the nonlinear acoustic waves propagate. From these principles a nonlinear acoustic wave equation (Kuznetsov's equation) is derived. The model equations for the nonlinear acoustic phenomena treated in detail in this book are Kuznetsov's equation and its approximations.

The books on nonlinear acoustics which have appeared after the book of Rudenko and Soluyan fall essentially into three categories: 1) Books concentrating on mathematics, e.g. nonlinear diffusion equations. 2) Books dealing with a wide range of applications. 3) Books dealing with special applications. A book of the first category is "Nonlinear Diffusive Waves" by P.L. Sachdev (Cambridge University Press, 1987). Books of the second category are "Nonlinear Acoustics", edited by M.F. Hamilton and D.T. Blackstock (Academic Press, San Diego, 1997) and "Nonlinear Wave Processes in Acoustics" by K. Naugolnykh and L. Ostrovsky (Cambridge University Press, 1998). Examples of books of the third category are "Nonlinear Theory of Sound Beams" by N.S. Bakhvalov, Ya. M. Zheleikin and E.A. Zabolotskaya (American Institute of Physics, New York, 1987), "Nonlinear Underwater Acoustics" by B.K. Novikov, O.V. Rudenko and V.I. Timoshenko (American Institute of Physics, New York, 1987) and "Nonlinear Random Waves and Turbulence in Nondispersive Media: Waves, Rays, Particles" by S.N. Gurbatov, A.N. Malakhov and A.I. Saichev (Manchester University Press, 1991).

The present book differs from books of the first category by its derivations of model equations from physical laws and principles and from books of the second category by its attempt to give a coherent theory of the fewer phenomena accounted for. On the other hand, the present book is less specialized than the books of the third category. It is our hope that the theoretical developments in this book will give some foundations for the study of wider range of applications accounted for in the books of categories 2 and 3.

We have consciously tried to avoid any geographical bias in our choice of references. Our list of references contains work of physical and mathematical character from most of the countries represented at the international symposia on nonlinear acoustics.

The present book has been written in close contact with our research and academic teaching since 1992. During this time we have had the advantage of discussing problems of nonlinear acoustics with many colleagues, from whom we have obtained information, advice and valuable points of view. We are especially grateful to D.T. Blackstock, the late D.G. Crighton, P.H. Enflo, S.N. Gurbatov, M.F. Hamilton, I. A. Molotkov, C. Nyberg, O.V. Rudenko, P.L. Sachdev and L.H. Söderholm. P.L. Sachdev has given many useful remarks and corrections to chapters 1-6 and O.V. Rudenko has, as our collaborator in scientific work, inspired the presentation in chapter 8.

Bengt O. Enflo
Claes M. Hedberg

Chapter 1

Introduction

1.1 The place of acoustics in fluid mechanics

Sound was originally the term for the sensation which stimulates the mechanism of the ear. The word acoustics, which denotes the science of sound, has its etymological origin in the Greek word *akouein*, which means hear. Already in the antiquity the idea that sound is wave motion similar to water waves was outspoken. The knowledge that a string, vibrating with a certain frequency, generates a vibratory air motion of the same frequency, corresponding to a musical note, dates back to Pythagoras (550 B.C.). However, the experimental evidence of this fact comes from the seventeenth century. From this century comes also the knowledge of sound as a wave motion in a medium, without which no sound propagation is possible.

In the eighteenth century, the theory of sound propagation was connected with basic theory of continuum physics. The pioneer was Leonhard Euler (1766). The wave equation, named after d'Alembert (1747), who derived its one-dimensional form for string motion, was found to play an important rôle both in fluid mechanics and elasticity. In sections 2.1 and 2.2 the connection between fluid mechanics and acoustics is demonstrated considering both nonlinear and dissipative effects, which had not yet been done in Euler's time.

Linear acoustics, modelled by d'Alembert's equation, has had an eminent success in describing most acoustical phenomena. Most basic and applied research in acoustics deals with sound generation, propagation, reflection, refraction and radiation, modelled as an initial or boundary value problem of d'Alembert's equation. However, since long time ago it has been known that there are acoustical wave propagation phenomena, which cannot be described by linear acoustics. In fact, the condition for linear acoustics to be applicable is that the amplitude of the sound wave is sufficiently small; finite amplitude sound must be described by nonlinear acoustics. Some important results in nonlinear acoustics were known already in the eighteenth century. However, nonlinear acoustics began its development as a specialty within the science of acoustics as late as in the middle of the twentieth century. In chapter 2 the theory of nonlinear acoustic wave motion is developed from the fundamental principles of the mechanics of continua.

1.2 Nonlinear acoustics before 1950

Some results on one important nonlinear acoustic phenomenon, large amplitude mechanical vibration, have been known since the eighteenth century. Euler (1766) has derived an equation for nonlinear plane acoustic waves in air assuming that the air had the properties of Boyle's law, which describes the behaviour of gas at constant temperature. Euler was also aware of the fact that the propagation speed of nonlinear waves differs from the one of linear waves. However, the correct law of propagation of plane progressive waves was not found until hundred years later by Earnshaw (1860). Progress in solving nonlinear plane wave equations were made by Lagrange (1761) and Poisson (1808) still considering waves in a gas obeying Boyle's law. The theory of nonlinear plane waves in a lossless fluid was much advanced by the works of G.B. Airy (1849), S. Earnshaw (1858, 1860) and B. Riemann (1860). The general properties of the Riemann wave equation are studied in section 3.1.

Another nonlinear acoustics phenomenon which was investigated early is shock formation. G.C. Stokes (1848) realized the difficulties associated with discontinuities of the nonlinear waves propagating in a lossless fluid. The description of the wave after the formation of the shock became an important problem. The main contribution here came from Rankine (1870) and Hugoniot (1887, 1889). They formulated conservation equations for mass, momentum and energy connecting the flow field behind a shock with the flow field ahead of it. An example of a result, derived by the use of the Rankine-Hugoniot relations (Rankine 1870), is that the weak shocks (i.e. the discontinuity of the field is small compared to the total field) travel with a speed, which is the mean value of the wave speeds just ahead of and behind the shock. This result is derived by using the mass conservation, in section 3.1. An early manifestation of nonlinear acoustics is the phenomenon in music called Tartini tones , discovered by the violinist Tartini in the eighteenth century. When two tones are sounded loudly and simultaneously on the violin, the tone with the difference frequency is heard. The nonlinear origin of combination tones, of which the difference frequency tone is an example, was realized by Helmholtz (1856). There is a vast literature on difference frequency tones (Rossing 1990), since they are important both for the perception of musical sound and for the understanding of the human auditory system.

In the theory of finite wave propagation in a nonviscous fluid substantial progress was made by Fubini-Ghiron (1935). Earnshaw (1858) gave an implicit solution to the nonlinear plane wave equation proposed earlier by Poisson (1808). Fubini's contribution was to find the Fourier expansion of that solution, before shock generation, in the case of an original sine wave. This result is discussed in section 4.2.

The need to include viscosity in the description of shock formation and propagation was realized already by Stokes (1848), but the first successful attempts were made by Lord Rayleigh (1910) and G.I. Taylor (1910). For weak shocks, Taylor obtained an expression for the shock profile (see section 5.1).

It is remarkable that a Fourier expansion of the nonlinear plane, originally monochromatic, wave in a slightly dissipative medium was found before the Fubini solution; it was given by Fay (1931). Fay's solution was obtained before Burgers' equation (Bateman 1915, Burgers

1948), the standard equation for plane waves in nonlinear acoustics, began to be used in acoustics. This means that the approximations made in the derivation of Burgers' equation from first principles are made in Fay's work. Fay's work is based on series expansions of the same kind as those used in section 5.2, where the expression found by Fay is proved to be an exact solution of Burgers' equation.

For the earlier history of acoustics reference is given to the book by Lindsay (1972). The earlier history of nonlinear acoustics, which is given here rather briefly, is presented with more details in an article by Blackstock (1997[a]) and in the book by Beyer (1985). The intime connections of nonlinear acoustics with other branches of physics, such as general relativity and quantum field theory, are discussed by Westervelt (1975).

1.3 Special phenomena in nonlinear acoustics

In this section the special nonlinear acoustic phenomena treated in this book are reviewed. These phenomena are studied by wave equations derived from first principles. Reviews of these equations, including equations for phenomena not treated in this book, such as dispersion and relaxation, are given by Rudenko, Soluyan and Khokhlov (1974), Crighton (1979) and in the book edited by M.F. Hamilton and D.T. Blackstock (1997). Wave equations studied in the present book are treated in that book in the articles by Hamilton and Morfey (1997), Blackstock, Hamilton and Pierce (1997) and Hamilton (1997).

1.3.1 Common theoretical description of nonlinear acoustics phenomena

In this book some nonlinear acoustical phenomena are studied, for which Riemann's and Burgers' equations do not give an adequate description. Therefore a more general wave equation, Kuznetsov's equation (Kuznetsov 1971), is derived from first principles. Kuznetsov's equation, which is derived in section 2.2, can be approximated to give

Riemann's equation, Burgers' equation and all other more realistic model equations for the phenomena studied. Kuznetsov's equation is a generalization of d'Alembert's equation with new terms due to nonlinearity and dissipation. The assumption of the relative smallness of these new terms leads directly, depending on geometry, to Burgers' equation (for plane waves) or to generalized Burgers' equations (for cylindrical and spherical waves), and, if dissipation is neglected, to Riemann's equation. Application of Kuznetsov's equation to sound beams leads to the Khokhlov - Zabolotskaya - Kuznetsov (KZK) equation (Zabolotskaya and Khokhlov 1969, Kuznetsov 1971). All these fundamental model equations of nonlinear acoustics are derived in section 2.3. These model equations describe the phenomena treated in chapters 3, 4, 5, 6 and 7, where dimensionless forms of these model equations are derived. Kuznetsov's equation in its original form is treated in chapter 8.

1.3.2 Generation and propagation of higher harmonics in travelling waves

After the achievements of Fay (1931) and Fubini-Ghiron (1935) nonlinear acoustics began to develop as a specialty in its own right, with its own methods different from those used in linear acoustics and other fields of fluid mechanics. The equation giving the most important results in the early days of nonlinear acoustics is due to J.M. Burgers. Its important of role was pointed out already by Bateman (1915). It was used by Burgers (1948) as a turbulence model. After the exact analytic solution of Burgers' equation was published (Hopf 1950, Cole 1951) it was used for nonlinear acoustic wave propagation in viscous fluids (Mendousse 1953) and in thermoviscous gases (Lighthill 1956). The exact analytic solution of Burgers' equation is presented in section 3.2.

A fundamental problem, on which substantial progress was made in the beginning of 60's, is the deformation and decay of a sine wave. The deformation is an accumulative process and increases when the wave proceeds. According to the Riemann equation the solution finally becomes multivalued, which requires the replacement of part of the solution by a discontinuity (shock). Before the appearance of the shock Fubini's solution is valid, giving the Fourier expansion of the deformed wave as a function of the distance from the boundary or as a function of the time after the preparation of a monochromatic wave in space. In the inviscid case the transition from the Fubini solution to the solution with shocks (sawtooth solution) was investigated by Blackstock (1966) and is discussed in section 4.2. When dissipation is considered, Fay's solution of Burgers' equation is used. It is a refinement of the sawtooth solution and valid after the shock formation. Cole (1951) derived Fay's solution as an asymptotic form of the exact solution of Burgers' equation evolving from a sine wave. Blackstock (1964) improved Fay's solution using Cole's (1951) exact solution of Burgers' equation. Another solution to Burgers' equation, which approximately evolves from an original sine wave and, like Fay's solution, is an exact solution to Burgers' equation, is given by Soluyan and Khokhlov (1961). An improvement by Blackstock of Fay's solution is detailed in section 5.2. The results by Soluyan and Khokhlov are given in section 5.3, where the agreement between Fay's and Khokhlov-Soluyan's solution is brought out.

Fay's solution, although an exact solution of Burgers' solution, is an approximate Fourier expansion of the solution evolving from a monochromatic wave. A recursion formula for the exact Fourier coefficients is given by Enflo and Hedberg (2001). These Fourier coefficients are numerically evaluated both in the pre-shock and in the shock regimes, thus providing a solution in the intermediate region, where neither Fubini's solution nor that of Fay is valid. These results, with smooth joining of the two solutions, are accounted for in section 5.4.

Because of the dissipation, the higher harmonics generated by nonlinearity are damped out, and the asymptotic wave is again monochromatic and satisfies Burgers' equation with the nonlinear term neglected. The asymptotic wave is identical with the first term in the Fourier series which constitutes Fay's solution. The wave amplitude in this term is found to be independent of the amplitude of the original monochromatic wave, a phenomenon called saturation.

Saturation occurs not only for plane waves described by Burgers' equation but also for waves in other geometries, described by generalized Burgers' equations. For determining asymptotic (also called old-age) solutions to these equations with a monochromatic boundary or initial value condition, we have to solve equations for which no analytical solutions are known. In this book cylindrical and spherical waves as described by generalized Burgers' equations are studied in chapter 6. The corresponding old-age problem for the original sine wave is studied by several authors (Shooter, Muir and Blackstock 1974; Crighton and Scott 1979; Scott 1981; Sachdev, Tikekar and Nair 1986; Sachdev and Nair 1989; Enflo 1996) and is accounted for in section 6.4.

1.3.3 Generation and propagation of combination frequency travelling waves

The cumulative process of deformation of a propagating wave has been studied for initial waves of a more complicated character than the monochromatic waves. The evolution of original bifrequency waves was studied by Fenlon (1972), who generalized the results to multifrequency sources, using the Riemann equation and obtained generalized Fubini solutions valid before the appearance of shocks. These results are given in section 4.3. Hedberg (1996, 1999) has also studied the evolution of multifrequency waves according to Burgers' equation and formulated an exact multifrequency solution, which includes dissipation. In section 5.5 this result of Hedberg is applied to an initial bifrequency wave and compared to the earlier results by Fenlon (1973^a). An exact solution generalizing the Mendousse solution was developed by Lardner (1982) for a bifrequency source.

One effect of multifrequency original signals is the appearance of difference frequencies. When using two high frequencies the particular low difference frequency features are narrow beamwidths, essentially no side lobes and a large bandwidth. This was first theoretically described by Westervelt (1963) and was named parametric array while experimental verification was made by Berktay (1965).

1.3.4 Propagation of travelling short pulses and N-waves

The weak shock profiles studied already by Taylor (1910) were later studied in the context of short pulses, using Riemann's and Burgers' equation and generalized Burgers' equations. Like periodic waves, which transform to sawtooth waves in the inviscid case, a single hump will transform into a triangular shock wave and a pulse with both a rarefaction and a compression will develop into an N-wave with a front shock and a rear shock. An extensive account for weak shock theory in the inviscid case is found in chapters 2 and 3 of the book by G. B. Whitham (1978). Short pulses and N-waves under the Riemann equation are studied in section 4.1.

An N-wave is created by the deformation, due to nonlinearity, of any rarefaction-compression pulse. Its evolution in plane and other geometries is of considerable interest. The asymptotic (old-age) form of an original N-wave is smooth in a model which takes into account dissipation. To calculate the amplitude of this old-age form is simple for plane waves, since an exact analytic solution of Burgers' equation exists. However, it is a difficult problem for cylindrical and spherical waves. Using Burgers' and generalized Burgers' equations several authors (Crighton and Scott 1979; Sachdev, Tikekar and Nair 1986; Hammerton and Crighton 1989) have studied the old-age problem for N-wave initial conditions analytically and numerically. For cylindrical N-waves B.O. Enflo (1998) has obtained the first two terms in a series for the old-age amplitude, giving a value of this amplitude in agreement with the numerical solution. The cylindrical N-wave problem has application to the shockwave from a supersonic vehicle (Whitham 1950, 1952).

The old-age problem for plane N-waves is solved in section 5.1 both with and without the use of the exact solution of Burgers' equation. The latter method can, in principle, be used for finding asymptotic forms N-waves evolving under different kinds of generalized Burgers' equations (Enflo 1998). In section 6.2 this method is used for finding the old-age waveform arising from an initial cylindrical N-wave. The results agree with the numerical results (Sachdev , Tikekar and

Nair 1986; Hammerton and Crighton 1989). This calculation of the cylindrical N-wave problem is applied to the decay of the shockwave from a supersonic vehicle in section 6.3. The results depend on the parameters of the medium and the shape and velocity of the vehicle.

1.3.5 Propagation of limited sound beams

Most of the analytical results in nonlinear acoustics concern propagation of waves in unlimited space. For limited sound beams a model equation was proposed in 1969 (Zabolotskaya and Khokhlov 1969). This equation was later generalized in order to account for dissipation (Kuznetsov 1971). This generalized beam equation is called the KZK equation. Solutions to this equation by successive approximations were developed (Bakhvalov, Zhileiki and Zabolotskaya 1987) and used for studying parametric arrays in underwater acoustics (Novikov, Rudenko and Timoshenko 1987; Hamilton 1997). A method for studying shocks in sound beams, which reduces the problem of solving the KZK equation to solving a generalized Burgers' equation (Sionoid 1993; Enflo 2000) is discussed in section 7.2.

1.3.6 Waves in closed tubes

Although they form part of nonlinear acoustics, standing nonlinear waves have not been treated in earlier books and reviews. The first important theoretical work on nonlinear waves in fluids in closed tubes started with the fundamental hydrodynamical equations (Chester 1964). For a disturbance which is basically a progressive wave, these equations have been discussed earlier (Lighthill 1956). By generalizing Lighthill's treatment, Chester showed that shock waves appear when a piston oscillates at one end of the tube with a frequency close to a resonant frequency of the tube. Chester's solution includes the effects of compressive viscosity and of shear viscosity in the boundary layer of the walls.

In section 8.1 the problem of standing waves in a closed tube is studied within the frame of linear theory and the condition for nonlinearity to be important is formulated. Boundary layer effects are neglected and the nonlinearity is assumed to be due to large vibration amplitudes of the fluid at resonance. In section 8.2 the equations of nonlinear standing waves are formulated, using Kuznetsov's equation and perturbation theory. Results which agree with and generalize Chester's results are derived. In section 8.3 results are derived for the wave field in the tube when dissipation is neglected and in section 8.4 when dissipation is considered.

Chapter 2

Physical theory of nonlinear acoustics

Sound is mechanical waves propagating in a medium. In order to understand the physics of sound it is necessary to study the general principles of motion of continuous media, which in the present book mean fluids. The characteristic features of acoustic wave motion are not apparent in the basic equations of fluid motion and fluid constitutive properties. Therefore these basic equations have to be simplified by elimination of all dependent variables except one, for which an acoustic wave equation is formulated. This procedure is completed in sections 2.1 and 2.2, resulting in a wave equation (Kuznetsov's equation), which can be specialized to describe the nonlinear acoustics phenomena studied in this book. From Kuznetsov's equation the equations of these phenomena are derived in section 2.3. Derivations of the equations of nonlinear acoustics are found in the book by Rudenko and Soluyan (1977) and in the review articles by Bjørnø (1976), Naze Tjøtta and Tjøtta (1981) and Coulouvrat (1992).

2.1 Basic theory of motion of a diffusive medium

The description of acoustic waves in a fluid is founded on the theory of motion of a fluid, which is considered to be continuous. The medium is diffusive, which means that in its description are included viscosity and heat conduction effects. The state of the fluid at a point $\mathbf{r} = (x_1, x_2, x_3)$ at time t is determined by six variables: the three state variables pressure (p), density (ρ) and temperature (T) and the three velocity components $\mathbf{v} = (v_1, v_2, v_3)$. The viscosity and the heat conduction coefficients, although in general functions of the state variables, are assumed to be constant. Thus, in order to solve the problem of finding the state of the medium at any time t at any point $\mathbf{r}$, the six functions $p(\mathbf{r}, t)$, $\rho(\mathbf{r}, t)$, $T(\mathbf{r}, t)$ and $v_i(\mathbf{r}, t)$ with $i = 1, 2, 3$, six equations are needed. Five of these equations are statements of the conservation theorems valid for every continuous medium, and the sixth one is a state equation motivated by the mechanical model of the fluid, looked upon as made up of molecules. The five conserved quantities are the mass, the three cartesian components of momentum and the energy. For the equation of the fluid, we assume the equation of an ideal gas, giving the pressure p as a function of the density ρ and the entropy per mass s. Using the state equation and the laws of thermodynamics we can then find how the temperature T, which occurs in the energy equation, depends on ρ and s, so the introduction of s does not mean that a new independent state variable has to be considered. Starting with these six equations the aim now is to deduce mathematical models for special important cases of propagation of acoustic waves with large enough amplitude that the linear acoustics theory cannot be used. These models are given by one nonlinear partial differential equation in which one of the variables velocity, pressure or density is the dependent variable, which is a function of space and time.

In this book the local values of the dependent variables are determined by space coordinates (x_1, x_2, x_3) (Eulerian coordinates) and time t. For many purposes the dependent variables are expressed in terms of the identification coordinates a_i for the specific particle under consideration (Lagrangian coordinates) and time t. Then the a_i

describe the particle position at the time $t = t_0$. The dependent variables given in Eulerian coordinates can be transformed to their values in Lagrangian coordinates by a Taylor expansion centred on the space coordinate x (Hunt 1955; Bjørnø 1976).

2.1.1 Conservation of mass; the continuity equation

Conservation of the mass of a fluid means that the net flow of mass due to the fluid motion from an arbitrary fixed volume V with the boundary S is equal to the reduction per time of the mass in the volume. The mathematical expresssion of this equality is

$$\int_S \rho v_k \, dS_k = -\frac{\partial}{\partial t} \int_V \rho \, dV \ , \qquad k = 1,2,3 \ , \tag{2.1}$$

where the summation convention

$$v_k \, dS_k \equiv \sum_{k=1}^{3} v_k \, dS_k \tag{2.2}$$

is used and will be used throughout. Transformation of the surface integral in (2.1) using Gauss' theorem gives

$$\int_V [\frac{\partial \rho}{\partial t} + \frac{\partial}{\partial x_i}(\rho v_i)] \, dV = 0 \ . \tag{2.3}$$

Since the volume V is arbitrary, the integrand in (2.3) must vanish:

$$\frac{\partial \rho}{\partial t} + \frac{\partial}{\partial x_i}(\rho v_i) = 0 \ . \tag{2.4}$$

Using the substantial time derivative

$$\frac{D}{Dt} \equiv \frac{\partial}{\partial t} + v_i \frac{\partial}{\partial x_i} \tag{2.5}$$

equation (2.4) becomes

$$\frac{D\rho}{Dt} + \rho \nabla \cdot \mathbf{v} = \mathbf{0} \ . \tag{2.6}$$

The equations (2.4) and (2.6) are the continuity equation written in two different manners. The continuity equation is the first one of the six fundamental equations necessary for deducing a theory of nonlinear acoustic wave propagation.

2.1.2 Conservation of momentum. Navier-Stokes equations

In order to formulate mathematically the conservation of momentum of a fluid we apply the same reasoning for a cartesian momentum density component ρv_i as we did for the density ρ when deriving the continuity equation (2.4). However, to the flow of momentum out of the fixed volume V shall be added the negative force exerted on the fluid in the volume V by volume and surface forces.

Thus we change (2.1) by the exchange $\rho \to \rho v_i$ and substract from the lefthand side the volume integral of the force per volume ρf_i and the surface integral of the symmetric stress tensor σ_{ik}:

$$\int_S \rho v_i v_k \, dS_k - \int_V \rho f_i \, dV - \int_S \sigma_{ik} \, dS_k = -\frac{\partial}{\partial t} \int_V \rho v_i \, dV \ . \qquad (2.7)$$

Writing the surface integrals in (2.7) as volume integrals by Gauss' theorem and using the arbitrariness of the volume V we can write (2.7) in differential form:

$$\frac{\partial}{\partial t}(\rho v_i) + \frac{\partial}{\partial x_k}(\rho v_i v_k) = \rho f_i + \frac{\partial}{\partial x_k}(\sigma_{ik}) \ . \qquad (2.8)$$

Using the continuity equation (2.4) in (2.8) we obtain Newton's second law of motion

$$\rho \frac{D v_i}{D t} \equiv \rho \frac{\partial v_i}{\partial t} + \rho v_{i,k} v_k = \rho f_i + \frac{\partial}{\partial x_k}(\sigma_{ik}) \ , \qquad (2.9)$$

where the substantial time derivative $\frac{Dv_i}{Dt}$ of the fluid velocity is the acceleration of a mass element.

For the construction of the stress tensor we have at our disposal the Kronecker tensor δ_{ik} and the deformation velocity tensor

$$d_{ik} \equiv \frac{1}{2}(v_{i,k} + v_{k,i}) \ , \qquad (2.10)$$

where $v_{i,k}$ is a shorter notation for $\frac{\partial v_i}{\partial x_k}$. We assume a linear dependence of σ_{ik} on the components of d_{ik} (Newtonian fluid):

$$\sigma_{ik} = -p(\rho, s)\delta_{ik} + \mu(v_{i,k} + v_{k,i} - \frac{2}{3}v_{r,r}\delta_{ik}) + \frac{2}{3}\lambda v_{r,r}\delta_{ik} \ , \tag{2.11}$$

where the viscosity coefficients λ and μ are assumed to be constants. Putting $f_k = 0$ and inserting (2.11) into (2.9) we obtain the Navier-Stokes equations (Navier 1823; Stokes 1845)

$$\rho(\frac{\partial v_i}{\partial t} + \frac{\partial v_i}{\partial x_k}v_k) = \frac{\partial p}{\partial x_i} + \mu\frac{\partial^2 v_i}{\partial x_k \partial x_k} + (\frac{2}{3}\lambda + \frac{\eta}{3})\frac{\partial}{\partial x_i}\frac{\partial v_k}{\partial x_k} \tag{2.12}$$

or alternatively, with $\mu = \eta$, $\frac{2}{3}\lambda = \zeta$:

$$\rho[\frac{\partial \mathbf{v}}{\partial t} + (\mathbf{v} \cdot \nabla)\mathbf{v}] = -\nabla p + \eta\Delta\mathbf{v} + (\zeta + \frac{\eta}{3})\text{grad div}\,\mathbf{v} \ . \tag{2.13}$$

The coefficient ζ is bulk viscosity and η is shear viscosity. The three equations (2.13) are among the six fundamental equations from which we will derive a theory of acoustic wave propagation in fluids. For the special case $\zeta = \eta = 0$ the equations (2.13) together with the continuity equation (2.6) are called Euler's fundamental hydrodynamic equations (Euler 1766).

2.1.3 Conservation of energy

In a way similar to what has been done for mass and momentum a conservation law for the energy can be formulated. This law says that the reduction per time of energy contents in a fixed volume V with boundary S has three contributions: the net internal energy flow from V through S because of fluid motion, the net heat flow out from V through S and the work per time done by the fluid in the volume on its surroundings. If u is the total energy per mass and q_i the heat flow density this law is expressed in a manner similar to (2.1):

$$\int_S \rho u v_i \, dS_i + \int_S q_i \, dS_i - \int_S v_i \sigma_{ik} \, dS_k = -\frac{\partial}{\partial t}\int_V \rho u \, dV \ . \tag{2.14}$$

The stress tensor is so defined that $\sigma_{ik}\, dS_k\, dx_i$ is the work done by the reaction force on the surface element $d\mathbf{S}$. Transforming the surface integrals in (2.14) by Gauss' theorem and using that V is an arbitrary volume we can write the differential form of the energy conservation theorem:

$$\frac{\partial}{\partial t}(\rho u) + \frac{\partial}{\partial x_i}(\rho u v_i) + \frac{\partial q_i}{\partial x_i} - \frac{\partial}{\partial x_k}(v_i \sigma_{ik}) = 0 \; . \tag{2.15}$$

We define σ'_{ik} as

$$\sigma'_{ik} = \eta(v_{i,k} + v_{k,i} - \frac{2}{3} v_{r,r}\delta_{ik}) + \zeta\, v_{r,r}\delta_{ik} \tag{2.16}$$

so that according to (2.11)

$$\sigma_{ik} = -p\delta_{ik} + \sigma'_{ik} \; . \tag{2.17}$$

Now (2.15) can be rewritten as:

$$\frac{\partial}{\partial t}(\rho u) + \frac{\partial}{\partial x_i}(\rho u v_i) = -\frac{\partial q_i}{\partial x_i} - \frac{\partial}{\partial x_i}(p v_i) + v_i \sigma'_{ik,k} + \Phi \; , \tag{2.18}$$

where

$$\Phi \equiv \sigma'_{ik} v_{i,k} = \frac{1}{2}\eta(v_{i,k} + v_{k,i} - \frac{2}{3} v_{r,r}\delta_{ik})^2 + \zeta(v_{k,k})^2 \; . \tag{2.19}$$

In (2.19) we have used the fact that only the symmetric and traceless part of $v_{i,k}$ contributes to the first term on the righthand side. The total energy u consists of internal energy $u^{(i)}$ and mechanical, i.e. kinetical and potential, energy $u^{(m)}$:

$$u = u^{(i)} + u^{(m)} \; . \tag{2.20}$$

An equation for the mechanical energy is obtained by multiplying (2.9) by v_i and assuming that f_i is conservative, i.e.

$$\rho f_i = -\rho \frac{\partial \Phi}{\partial x_i} \; . \tag{2.21}$$

In this way we obtain

$$\rho v_i \frac{Dv_i}{Dt} = \rho v_i \frac{\partial \Phi}{\partial x_i} + v_i \frac{\partial \sigma_{ik}}{\partial x_k} , \tag{2.22}$$

or, using (2.4), (2.5) and (2.17):

$$\frac{\partial}{\partial t}(\frac{1}{2}\rho \mathbf{v}^2 + \rho\Phi) + \frac{\partial}{\partial x_i}[\rho v_i(\frac{1}{2}\mathbf{v}^2 + \Phi)] = v_i \frac{\partial \sigma'_{ik}}{\partial x_k} - v_i \frac{\partial p}{\partial x_i} . \tag{2.23}$$

Putting

$$u^{(m)} = \frac{1}{2}\mathbf{v}^2 + \Phi \tag{2.24}$$

and subtracting (2.23) from (2.18) we obtain

$$\frac{\partial}{\partial t}(\rho u^{(i)}) + \frac{\partial}{\partial x_k}(\rho u^{(i)} v_k) = -\frac{\partial q_k}{\partial x_k} - p\nabla \cdot \mathbf{v} + \Phi . \tag{2.25}$$

where we have used (2.20). We assume that the heat flow density q_k is proportional to the temperature gradient $\frac{\partial T}{\partial x_k}$ with the proportionality factor $-\kappa$, which is assumed to be a constant:

$$q_k = -\kappa \frac{\partial T}{\partial x_k} , \tag{2.26}$$

where κ is called the heat conduction number. Combination of the first and second laws of thermodynamics gives the wellknown relation:

$$du^{(i)} = T ds - p\, d(\frac{1}{\rho}) , \tag{2.27}$$

where s is entropy per mass and $\frac{1}{\rho}$ is the specific volume. Using the continuity equation (2.6) we obtain from (2.27) by substantial time derivation:

$$\rho \frac{Du^{(i)}}{Dt} = \rho T \frac{Ds}{Dt} - p\nabla \cdot \mathbf{v} . \tag{2.28}$$

From the continuity equation (2.4) it is easy to see that the following relation is valid for an arbitrary variable, for example $u^{(i)}$:

$$\rho \frac{Du^{(i)}}{Dt} = \frac{\partial}{\partial t}(\rho u^{(i)}) + \frac{\partial}{\partial x_k}(\rho u^{(i)} v_k) . \tag{2.29}$$

Thus the lefthand sides of the equations (2.25) and (2.28) are equal. Putting the righthand sides equal in the same equations and using (2.26) and (2.19) we obtain

$$\rho T[\frac{\partial s}{\partial t} + (\mathbf{v} \cdot \nabla)s] = \kappa \Delta T + \zeta(\nabla \cdot \mathbf{v})^2 + \frac{1}{2}\eta(\partial_i v_j + \partial_j v_i - \frac{2}{3}\nabla \cdot \mathbf{v}\delta_{ij})^2 \ . \tag{2.30}$$

This form of the energy conservation theorem will be useful later on. It is referred to by Pierce (1981) as the Kirchhoff-Fourier equation (Kirchhoff 1868; Fourier 1822). As Pierce points out, this terminology is somewhat inaccurate, since neither Kirchhoff nor Fourier used the concept of entropy in their relevant publications. With (2.6), (2.13) and (2.30) all conservation theorems that are needed have been brought into the required form.

2.1.4 Ideal fluid equation of state

We assume the equation of state of an ideal fluid to be valid for the medium:

$$\frac{p}{\rho} = (c_p - c_v)T \ , \tag{2.31}$$

where c_p, c_v are the capacities per unit mass of the fluid of constant pressure and volume respectively. It is accounted for by Pierce (1981) how this equation is derived from Boyles's law and the definition of absolute temperature.

It is convenient to introduce the enthalpy per unit mass h:

$$h = u^{(i)} + \frac{p}{\rho} \ . \tag{2.32}$$

Thus by comparing (2.27) and (2.32) we find

$$dh = Tds + \frac{1}{\rho}dp = dq + \frac{1}{\rho}dp \ , \tag{2.33}$$

where dq is the heat per unit mass brought to the fluid. By the defintition of c_p we obtain from (2.33)

$$c_p = \left(\frac{\partial h}{\partial T}\right)_p \ . \tag{2.34}$$

Using the mathematical identity

$$\left(\frac{\partial T}{\partial p}\right)_h = -\frac{(\frac{\partial h}{\partial p})_T*}{(\frac{\partial h}{\partial T})_p*} , \tag{2.35}$$

(2.33), (2.34) and the Maxwell relation

$$\left(\frac{\partial s}{\partial p}\right)_T = -\left(\frac{\partial \frac{1}{\rho}}{\partial T}\right)_p \tag{2.36}$$

we obtain

$$\begin{aligned}\left(\frac{\partial T}{\partial p}\right)_h &= -\frac{(\frac{\partial h}{\partial p})_T*}{(\frac{\partial h}{\partial T})_p*} = -\frac{1}{c_p}\left(T\left(\frac{\partial s}{\partial p}\right)_T + \frac{1}{\rho}\right) \\ &= \frac{1}{c_p}\left(T\left(\frac{\partial \frac{1}{\rho}}{\partial T}\right)_p - \frac{1}{\rho}\right) = 0 .\end{aligned} \tag{2.37}$$

where (2.31) has been taken into account in the last equality. Consequently we obtain from (2.37) and (2.35)

$$\left(\frac{\partial h}{\partial p}\right)_T = 0 , \tag{2.38}$$

and from (2.34) and (2.38), with a suitable choice of integration constant,

$$h = c_p T . \tag{2.39}$$

It is also of interest to express, for an ideal fluid, the total differential of the enthalpy with s and ρ as independent variables. Inserting (2.31) into (2.33) we find, with (2.39),

$$dh = Tds + \frac{1}{\rho}(c_p - c_v)\left(\frac{h}{c_p}d\rho + \frac{\rho}{c_p}dh\right) \tag{2.40}$$

and thus

$$dh = \gamma T ds + (\gamma - 1)h\frac{d\rho}{\rho} , \tag{2.41}$$

where

$$\gamma = \frac{c_p}{c_v} . \tag{2.42}$$

2.2 Derivation of the three dimensional wave equation of nonlinear acoustics (Kuznetsov's equation)

The aim of this section is to eliminate all dependent variables except one from the system (2.6), (2.13), (2.30) and (2.31) and deduce a nonlinear wave equation for that variable. In order to simplify the analysis we assume that the flow is irrotational. This implies

$$\nabla \times \mathbf{v} = 0 \tag{2.43}$$

$$\mathbf{v} = -\nabla\Phi \ , \tag{2.44}$$

where Φ is the velocity potential.

By means of (2.43), (2.44) the relations (2.13) can be simplified. The simplification is made using the operator relation (grad $\equiv \nabla$, div $\equiv \nabla\cdot$, rot $\equiv \nabla\times$)

$$\text{rot rot} = \text{grad div} - \Delta \tag{2.45}$$

and the relation

$$\frac{1}{2}\nabla\mathbf{v}^2 = (\mathbf{v}\cdot\nabla)\mathbf{v} + \mathbf{v}\times(\nabla\times\mathbf{v}) \ . \tag{2.46}$$

Using (2.44) - (2.46) together with (2.13) a simpler form of Navier-Stokes' equations is found:

$$\nabla[-\frac{\partial\Phi}{\partial t} + \frac{1}{2}(\nabla\Phi)^2] = -\frac{1}{\rho}\nabla p - \frac{1}{\rho}(\frac{4}{3}\eta + \zeta)\nabla(\Delta\Phi) \ . \tag{2.47}$$

The deviations of ρ, p, s and T from their equilibrium values ρ_0*, p_0, s_0 and T_0 are assumed to be small, as well as the fluid velocity $|\mathbf{v}|$. The heat conduction number κ and the viscocities η, ζ will also be treated as small quantities. Of course these assumptions of smallness correspond to the assumption of some dimensionless numbers being much less than unity. We do not make the analysis lengthier by introducing these numbers but just state that all equations are written retaining the second order of the small quantities. Then (2.30) simplifies to

$$\rho_0 T_0 \frac{\partial s}{\partial t} = \kappa \Delta T \tag{2.48}$$

because the neglected terms are of the third order of small quantities. Since the righthand side of (2.48) is small of the second order, the equation (2.48) shows that the deviation of the entropy from its equilibrium value must be treated as small of the second order. In the replacement of ρ and T by their equilibrium values at the lefthand side of (2.48) only third order contributions are neglected.

In order to obtain a wave equation we must eliminate the term $\frac{1}{\rho}\nabla p$ in (2.47). We follow a procedure given by Söderholm (2001) [1] and obtain after using (2.33):

$$\frac{1}{\rho}\nabla p = \nabla h - T_0 \nabla s \ . \tag{2.49}$$

Because entropy changes are small of the second order we have replaced T by T_0 on the righthand side of (2.49). Inserting (2.49) into (2.47) we can integrate the latter equation and obtain:

$$-\frac{\partial \Phi}{\partial t} + \frac{1}{2}(\nabla \Phi)^2 + h + T_0(s - s_0) + \frac{1}{\rho}(\frac{4}{3}\eta + \zeta)\Delta \Phi = c_p T_0 \ . \tag{2.50}$$

The second term on the lefthand side of (2.47) is of second order so in (2.50) we have replaced ρ by $\rho_0{}^*$ and considered the small quantities η, ζ as constants. The integration constant on the righthand side of (2.50) is obtained from (2.39).

The entropy will now be eliminated from (2.50) by use of (2.48), which we write using (2.39):

$$\rho_0 T_0 ** \frac{\partial s}{\partial t} = \frac{\kappa}{c_p} \Delta h \ . \tag{2.51}$$

Since (2.51) is of the second order, we use in (2.51) the first order expression for h obtained from (2.50):

$$h \approx c_p T_0 + \frac{\partial \Phi}{\partial t} \ . \tag{2.52}$$

---

[1] A similar procedure, neglecting heat conduction, is given by Campos (1986).

Integration of (2.51) then gives:

$$T_0(s - s_0) = \frac{\kappa}{\rho_0 c_p} \Delta\Phi \ . \tag{2.53}$$

Using the result from (2.53) in (2.50) and deriving with respect to t we obtain

$$-\frac{\partial^2\Phi}{\partial t^2} + (\nabla\Phi) \cdot \nabla\frac{\partial\Phi}{\partial t} + \frac{\partial h}{\partial t} + \frac{1}{\rho}(-\frac{\kappa}{c_p} + \frac{4}{3}\eta + \zeta)\Delta\frac{\partial\Phi}{\partial t} = 0 \ . \tag{2.54}$$

We finally have to eliminate the enthalpy h in (2.54). From (2.41) and the continuity equation (2.6) we obtain the following equation correct to the second order:

$$\frac{D}{Dt}[h - \gamma T_0(s - s_0)] - (\gamma - 1)h \, \Delta\Phi = 0 \ . \tag{2.55}$$

Combining (2.53) and (2.55) gives the following equation correct to the second order:

$$\frac{\partial h}{\partial t} - \nabla\frac{\partial\Phi}{\partial t} \cdot \nabla\Phi - \frac{\kappa\gamma}{\rho_0 c_p}\Delta\frac{\partial\Phi}{\partial t} - (\gamma - 1)(c_p T_0 + \frac{\partial\Phi}{\partial t})\Delta\Phi = 0 \ . \tag{2.56}$$

In (2.56) we have used the first order expression (2.52) for h in those terms where a second order expression for h would give contributions of the third order.

The sound velocity c_0 in a fluid is given as

$$c_0^2 = [(\frac{\partial p}{\partial \rho})_s]_{\rho=\rho_0} \ , \tag{2.57}$$

(see e.g. Pierce (1981)). The state equation of an ideal fluid for adiabatical changes is Poisson's equation

$$\frac{p}{p_0} = (\frac{\rho}{\rho_0})^\gamma \tag{2.58}$$

The undisturbed sound velocity c_0 is given by (2.57) and (2.58),

$$c_0^2 = \gamma\frac{p_0}{\rho_0} \ . \tag{2.59}$$

Using (2.59) and eliminating $\frac{\partial h}{\partial t}$ in (2.54) by (2.56) we obtain

$$\frac{\partial^2 \Phi}{\partial t^2} - c_0^2 \Delta \Phi = \frac{\partial}{\partial t}[(\text{grad}\,\Phi)^2 + \frac{1}{2c_0^2}(\gamma - 1)\left(\frac{\partial \Phi}{\partial t}\right)^2 + \frac{b}{\rho_0}\Delta\Phi] \,, \quad (2.60)$$

where b is defined as

$$b = \kappa\left(\frac{1}{c_v} - \frac{1}{c_p}\right) + \frac{4}{3}\eta + \zeta \,. \quad (2.61)$$

Thus, to the second order in the small quantities, the total effect of viscosity and heat conduction is given by the single constant b. This equation was derived by V. P. Kuznetsov (1971) and gives a method of treating nonlinear and dissipative effects in all acoustic wave problems which, in the lowest approximation, are treated by d'Alembert's equation, i.e. by neglecting the second order terms on the righthand side of (2.60). In this book Kuznetsov's equation is used for the derivation of a number of nonlinear model wave equations, motivated by different geometrical conditions and assumptions of diffusivity. [2]

The number $\gamma - 1$ in the coefficient of the second term of the righthand side of Kuznetsov's equation (2.60) is called the parameter of nonlinearity of the (ideal) fluid. Following Beyer (1959, 1997) we will give the physical background of this assignment.

The pressure p of the fluid is considered as a function $p(\rho, s)$, given by the equation of state, of the density ρ and the entropy per mass s. We make a Taylor series expansion of the deviation $p - p_0$ from the equilibrium pressure value p_0 in the corresponding density deviation $\rho - \rho_0$ at constant entropy $s = s_0$:

$$p - p_0 = [(\frac{\partial p}{\partial \rho})_s]_{\rho=\rho_0}(\rho - \rho_0) + \frac{1}{2!}[(\frac{\partial^2 p}{\partial \rho^2})_s]_{\rho=\rho_0}(\rho - \rho_0)^2 + \dots \,. \quad (2.62)$$

The Taylor expansion (2.62) can be written in an equivalent manner, introducing the quantities A and B,

$$p - p_0 = A\frac{\rho - \rho_0}{\rho_0} + \frac{B}{2!}\left(\frac{\rho - \rho_0}{\rho_0}\right)^2 + \dots \,, \quad (2.63)$$

[2] A derivation of Kuznetsov's equation considering cubic terms in Φ is given by Söderholm (2001).

where A and B are defined as

$$A = \rho_0 [(\frac{\partial p}{\partial \rho})_s]_{\rho=\rho_0} \tag{2.64}$$

$$B = \rho_0^2 [(\frac{\partial^2 p}{\partial \rho^2})_s]_{\rho=\rho_0} \ . \tag{2.65}$$

Equations (2.64) and (2.65) together with (2.58) give for an ideal fluid:

$$\frac{B}{A} = \gamma - 1 \ . \tag{2.66}$$

The ratio B/A and its importance as a measure of wave distortion in liquids was first discussed by Fox and Wallace (1954). Several methods have been used to determine B/A for different kinds of liquids, gases and biological tissues. (Coppens, Beyer, Seiden, Donohue, Guepin, Hodson and Townsend 1965; Beyer 1959, 1997; Bjørnø 1986; Cobb 1983). The effect of third-order terms in (2.63) is discussed by Van Buren and Breazeale (1968).

2.3 Wave equations of nonlinear acoustics

In this section we shall, from Kuznetsov's equation (2.60), derive some of the most important wave equations in nonlinear acoustics. They are Burgers' equation describing plane waves in homogeneous space, generalized Burgers' equations describing cylindrical and spherical waves in homogeneous space and the Khoklov-Zabolotskaya-Kuznetsov equation describing plane waves in bounded beams.

2.3.1 Burgers' equation

We begin with deriving from (2.60) an equation for plane waves in homogeneous space. It is assumed that the wave profile, i.e. the time dependence of the wave at a given distance from the sound source, changes slowly with space because of the nonlinear and dissipative terms of the righthand side of (2.60). Without that righthand side there would be no change at all of the wave profile. The assumption of slow change of the wave profile thus is founded on the smallness of the righthand side of (2.60) and can, more precisely, be formulated as the assumption that a substantial change of the wave profile requires that the wave travels many wavelenghts. This means that the space derivative of the wave function is a small quantity. This smallness is made visible by rescaling the space variable by a small parameter μ. Because a disturbance passing the point $x = 0$ at the time $t = 0$ has moved to the point $x = c_0 t$ at the time t, the development of the waveform is best studied using a retarded time, subtracting the travel time x/c_0. The independent variables (x, t) are replaced by (x', τ) defined as

$$\begin{aligned} x' &= \mu x \\ \tau &= t - \frac{x}{c_0} \,. \end{aligned} \tag{2.67}$$

Consequently we obtain

$$\begin{aligned} \frac{\partial}{\partial x} &= -\frac{1}{c_0}\frac{\partial}{\partial \tau} + \mu \frac{\partial}{\partial x'} \\ \frac{\partial}{\partial t} &= \frac{\partial}{\partial \tau} \,. \end{aligned} \tag{2.68}$$

When inserting (2.67) and (2.68) into (2.60), considering μ, b and Φ as small and retaining only second order terms the result becomes

$$2\mu c_0 \frac{\partial^2 \Phi}{\partial \tau \partial x'} = \frac{\partial}{\partial \tau}\Big[\frac{1}{c_0^2}\Big(\frac{\partial \Phi}{\partial \tau}\Big)^2 + \frac{b}{\rho_0 c_0^2}\frac{\partial^2 \Phi}{\partial \tau^2} + \frac{\gamma - 1}{2c_0^2}\Big(\frac{\partial \Phi}{\partial \tau}\Big)^2\Big] \,. \tag{2.69}$$

We now return to the fluid velocity (cf. (2.44))

$$v = -\frac{\partial \Phi}{\partial x} = \frac{1}{c_0}\frac{\partial \Phi(x, \tau)}{\partial \tau} + o(\mu) \tag{2.70}$$

as the dependent variable in (2.69). Using x instead of x' we obtain in the lowest order from (2.69) and (2.70):

$$\frac{\partial v}{\partial x} - \frac{\beta}{c_0^2} v \frac{\partial v}{\partial \tau} - \frac{b}{2c_0^3 \rho_0}\frac{\partial^2 v}{\partial \tau^2} = 0 \tag{2.71}$$

where

$$\beta = \frac{\gamma + 1}{2} \ . \tag{2.72}$$

Equation (2.71) is *Burgers' equation*, proposed in 1948 by J. M. Burgers (1948, 1974) as a model for turbulence. It describes plane wave propagation in a diffusive medium when neither dissipation nor nonlinear effects can be neglected. It can be solved for a boundary condition which can be realized, for example, by a piston movement with a prescribed time dependence at the end of a long pipe filled with a homogeneous fluid.

It is practical to introduce dimensionless variables, using a characteristic velocity v_0 of the medium and a characteristic time $1/\omega$.

$$V = \frac{v}{v_0} \tag{2.73}$$

$$\theta = \omega\tau \tag{2.74}$$

$$\sigma = \frac{\beta}{c_0^2}\omega v_0 x \ . \tag{2.75}$$

Insertion of (2.73)-(2.75) into Burgers' equation (2.71) gives the dimensionless equation

$$\frac{\partial V}{\partial \sigma} - V\frac{\partial V}{\partial \theta} - \epsilon\frac{\partial^2 V}{\partial \theta^2} = 0 \ , \tag{2.76}$$

where

$$\epsilon = \frac{1}{2\beta}\frac{b\omega}{c_0 v_0 \rho_0} = \frac{1}{2\beta \mathsf{Re}} \ . \tag{2.77}$$

The number Re is called the *acoustical Reynolds number*. If $\mathsf{Re} \geq 1$, linear acoustic theory cannot be used, but must be replaced by nonlinear acoustic theory. In this book we assume nonlinearity and in general $\mathsf{Re} \gg 1$.

It must be pointed out that Burgers' equation can also be used as a model for a nonlinear plane wave experiment of another type than that modelled by (2.71). Instead of a boundary condition it is possible to prescribe an initial condition, i.e. a given x-dependent state of the fluid at a given time. In this case we do not make the change of variable (2.67), but instead

$$
\begin{aligned}
t' &= \mu t \\
\xi &= x - c_0 t \ .
\end{aligned}
\tag{2.78}
$$

From (2.60) the transformation (2.78) leads to the lowest order of Burgers' equation for $v(\xi, t)$:

$$
\frac{\partial v}{\partial t} + \beta v \frac{\partial v}{\partial \xi} - \frac{b}{2\rho_0} \frac{\partial^2 v}{\partial \xi^2} = 0 \ . \tag{2.79}
$$

2.3.2 Generalized Burgers' equation

We will now derive wave equations for cylindrical and spherical waves from (2.60). To that end is introduced the variable r, which is the radial variable in cylindrical or spherical coordinates. We assume that Φ depends only on r and t. In this case we have

$$
(\nabla \Phi)^2 = \left(\frac{\partial \Phi}{\partial r}\right)^2 \tag{2.80}
$$

$$
\Delta \Phi = \frac{\partial^2 \Phi}{\partial r^2} + \frac{n}{r} \frac{\partial \Phi}{\partial r} \ , \tag{2.81}
$$

where $n = 1$ for cylindrical waves and $n = 2$ for spherical waves. Both for spherical and cylindrical waves we introduce the new variables

$$
\begin{aligned}
r' &= \mu r \\
\tau &= t \mp \frac{r - r_0}{c_0} \ ,
\end{aligned}
\tag{2.82}
$$

where the minus (plus) sign is valid for diverging (converging) waves and r_0 is the radius of the cylindrical or spherical sound source (sound absorbing area). Inserting (2.80)-(2.82) into (2.60) and retaining second order terms we obtain for diverging waves the equation

$$
2c_0 \frac{\partial^2 \Phi}{\partial \tau \partial r} + \frac{n c_0}{r} \frac{\partial \Phi}{\partial \tau} = (\gamma + 1) \frac{\partial \Phi}{\partial \tau} \frac{\partial^2 \Phi}{\partial \tau^2} + \frac{b}{\rho_0 c_0^2} \frac{\partial^3 \Phi}{\partial \tau^3} \ . \tag{2.83}
$$

Since the fluid velocity is radial, we define

$$v \equiv v_r = -\frac{\partial \Phi}{\partial r} = \frac{1}{c_0}\frac{\partial \Phi}{\partial \tau} + o(\mu) \tag{2.84}$$

and obtain

$$\frac{\partial v}{\partial r} + \frac{n}{2r}v - \frac{\beta}{c_0^2}v\frac{\partial v}{\partial \tau} = \frac{b}{2c_0^3\rho_0}\frac{\partial^2 v}{\partial \tau^2}\,, \tag{2.85}$$

which is the generalized Burgers' equation for cylindrical ($n = 1$) and spherical ($n = 2$) waves, suitable for a boundary value problem (Naugolnykh, Soluyan and Khokhlov 1963[a]).

A wave equation suitable for an initial value problem can be found if instead of r' and τ given by (2.82) the following new variables are introduced:

$$\begin{aligned} t' &= \mu t \\ \rho &= r - r_0 \mp c_0 t \end{aligned} \tag{2.86}$$

A calculation similar to that leading to (2.85) now gives the generalized Burgers' equation for cylindrical ($n = 1$) and spherical ($n = 2$) diverging waves describing their time evolution for a given initial space dependence:

$$\frac{\partial v}{\partial t} + \frac{n}{2t}v + \beta v\frac{\partial v}{\partial \rho} = \frac{b}{2\rho_0}\frac{\partial^2 v}{\partial \rho^2}\,. \tag{2.87}$$

2.3.3 The KZK equation

Kuznetsov's equation (2.60) can also be used as a model for acoustic waves in limited volumes. One very important such problem, leading to a rather simple wave equation, is the propagation of a sound beam. We assume that the sound field is described by the function $\Phi(\mu x, \sqrt{\mu}\, y, \sqrt{\mu}\, z, \tau = t - \frac{x}{c_0})$, where μ is a small number. This means that a wave propagating in the x-direction is studied and that the change of the wave field in the transverse direction is faster than that in the longitudinal direction. Changing to the variables

$$\begin{aligned} x' &= \mu x \\ y' &= \sqrt{\mu}\, y \end{aligned}$$

$$z' = \sqrt{\mu}\, z$$
$$\tau = t - \frac{x}{c_0} \tag{2.88}$$

in (2.60) and retaining the second order terms in the small quantities μ, Φ and b we obtain, after returning to the variables x, y, z:

$$\frac{\partial}{\partial \tau}[\frac{\beta}{c_0^2}(\frac{\partial \Phi}{\partial \tau})^2 + \frac{b}{\rho_0 c_0^2}\frac{\partial^2 \Phi}{\partial \tau^2} - 2c_0 \frac{\partial^2 \Phi}{\partial x^2}] = -c_0^2(\frac{\partial^2 \Phi}{\partial y^2} + \frac{\partial^2 \Phi}{\partial z^2}) . \tag{2.89}$$

Putting

$$v \equiv v_x = -\frac{\partial \Phi}{\partial x} = \frac{1}{c_0}\frac{\partial \Phi}{\partial \tau} + O(\mu^2) \tag{2.90}$$

we obtain from (2.89) the *Khokhlov-Zabolotskaya-Kuznetsov (KZK) equation*

$$\frac{\partial}{\partial \tau}[\frac{\partial v}{\partial x} - \frac{\beta}{c_0^2} v \frac{\partial v}{\partial \tau} - \frac{b}{2\rho_0 c_0^3}\frac{\partial^2 v}{\partial \tau^2}] = \frac{c_0}{2}(\frac{\partial^2 v}{\partial y^2} + \frac{\partial^2 v}{\partial z^2}) \tag{2.91}$$

describing the sound field in a nonlinear bounded beam (Zabolotskaya and Khokhlov 1969; Kuznetsov 1971).

The form (2.91) of the KZK equation is suitable for boundary value problems. Like for the different forms of Burgers' equation we will give a form of the KZK equation suitable for initial value problems. Making the substitutions

$$\xi = x - c_0 t \tag{2.92}$$
$$y' = \sqrt{\mu} y \tag{2.93}$$
$$z' = \sqrt{\mu} z \tag{2.94}$$
$$t' = \mu t \tag{2.95}$$

and putting (remember that Φ is $O(\mu)$)

$$v \equiv v_x = -\frac{\partial \Phi}{\partial \xi} + O(\mu^2) \tag{2.96}$$

we obtain

$$\frac{\partial}{\partial \xi}[\frac{\partial v}{\partial t} + \beta v \frac{\partial v}{\partial \xi} - \frac{b}{2\rho_0}\frac{\partial^2 v}{\partial \xi^2}] = -\frac{c_0}{2}(\frac{\partial^2 v}{\partial y^2} + \frac{\partial^2 v}{\partial z^2}) . \tag{2.97}$$

Chapter 3

Basic methods of nonlinear acoustics

The wave equations of nonlinear acoustics derived in the preceding chapter cannot be solved exactly in general. However, Burgers' equation (2.71) and (2.79) has an exact solution and consequently also the special case of Burgers' equation with vanishing dissipation, called the Riemann equation. The general solution methods of Riemann's and Burgers' equations are treated in this chapter.

3.1 Solution methods to the Riemann wave equation

3.1.1 Physical interpretation of the Riemann equation

The simplest model of nonlinear acoustic wave propagation is obtained from (2.71) or (2.79) neglecting the dissipative effects by putting $b = 0$. If this simplification is introduced in Navier-Stokes' equations (2.13) we obtain Euler's hydrodynamic equations. In the middle of the 19th century Riemann (1860) developed methods for general solution of these nonlinear equations, using the concepts now known as characteristics and Riemann invariants. For this reason the waves studied in this chapter are called Riemann waves.

We start with the initial value condition form of Burgers' equation (2.79) with $b = 0$,

$$\frac{\partial v}{\partial t} + \beta v \frac{\partial v}{\partial \xi} = 0 \; . \qquad (3.1)$$

The equation (3.1) is called the Riemann equation. In order to make its physical interpretation clearer we write it using the variables $x = c_0 t + \xi$ and t, (cf. (2.78)),

$$\frac{\partial v}{\partial t} + (c_0 + \beta v)\frac{\partial v}{\partial x} = 0 \; . \qquad (3.2)$$

The total time derivative of v is

$$\frac{dv}{dt} = \frac{\partial v}{\partial t} + \frac{\partial v}{\partial x}\frac{dx}{dt} \; . \qquad (3.3)$$

We see, by comparing (3.2) and (3.3), that the velocity v is constant on the curves in the (x, t)-plane given by their derivatives

$$\frac{dx}{dt} = c_0 + \beta v(x, t) \; . \qquad (3.4)$$

These curves are called characteristic curves. The Riemann invariant (v) is constant on these curves, which are straight lines, as is obvious from (3.4). The concepts characteristic curves and Riemann invariants are treated more extensively in the book by Whitham (1974) in connection with systems of nonlinear equations in gas dynamics.

We will now see that the plane progressive wave velocity in a non-diffusive nonlinear fluid is the sum of the sound speed c and the particle velocity v. To that end we write the ratio c^2/c_0^2, where c is the sound speed in the disturbed fluid,

$$\frac{c^2}{c_0^2} = \frac{(\frac{\partial p}{\partial \rho})_s}{[(\frac{\partial p}{\partial \rho})_s]_{\rho=\rho_0}} = \frac{[(\frac{\partial p}{\partial \rho})_s]_{\rho=\rho_0} + (\rho - \rho_0)[(\frac{\partial^2 p}{\partial \rho^2})_s]_{\rho=\rho_0} + \cdots}{[(\frac{\partial p}{\partial \rho})_s]_{\rho=\rho_0}} \qquad (3.5)$$

$$= 1 + \frac{B}{A}\frac{\rho - \rho_0}{\rho_0} + \cdots \; , \qquad (3.6)$$

where we have used (2.64) and (2.65). Taking the square root of (3.6), performing a binomial expansion and using the linear relation (Pierce 1981)

$$\frac{\rho - \rho_0}{\rho_0} = \frac{v}{c_0} \tag{3.7}$$

yields

$$c = c_0 + \frac{B}{2A} v = c_0 + \frac{\gamma - 1}{2} v \ , \tag{3.8}$$

where (2.66) is used. Using (3.8) and (2.72), the propagation velocity $\frac{dx}{dt}|_v$, belonging to the particle velocity v (cf. (3.4)), of a plane progressive wave is obtained as

$$\frac{dx}{dt}|_v = c_0 + \beta v = c + v \ , \tag{3.9}$$

which is the presupposed result.

3.1.2 Continuous wave solution

Since the following considerations are of a little more general character, we change notation and notice that equation (3.1) is a special case of the general nonlinear equation

$$\rho_t + c(\rho)\rho_x = 0 \ , \tag{3.10}$$

where ρ_t, ρ_x mean partial derivatives and $c(\rho)$ is an arbitrary function of ρ. The equation (3.10) can be integrated if its lefthand side is a derivative with respect to either x or t. This is the case if the derivative is taken along certain curves in the (x, t) plane. The derivative of ρ with respect to t along the curve $x(t)$ is:

$$\frac{d\rho(x,t)}{dt} = \frac{\partial \rho}{\partial x}\frac{dx}{dt} + \frac{\partial \rho}{\partial t} \ . \tag{3.11}$$

Comparing Eqs. (3.10) and (3.11) we conclude that along every curve in the (x, t) plane with the derivative

$$\frac{dx}{dt} = c(\rho(x,t)) \tag{3.12}$$

the lefthand side of the equation (3.10) is a total derivative. Let us call a curve satisfying (3.10) for C. On the curve C the following relations are valid:

$$\frac{d\rho}{dt} = 0 \tag{3.13}$$

$$\frac{dx}{dt} = c(\rho) . \tag{3.14}$$

According to (3.13) ρ is constant on C. The equation (3.14) then tells us that dx/dt, i.e. the slope of the curve C, is constant on C and therefore the curves C are straight lines in the (x, t) plane. We now prescribe the initial condition

$$\rho = f(x) \qquad \text{for } t = 0, \ -\infty < x < \infty . \tag{3.15}$$

Assume that a curve C crosses the x axis $t = 0$ at the point $x = \xi$. Since ρ is constant on the curve C, the constant ρ value must be $\rho = f(\xi)$. The slope of the curve is $c(f(\xi))$, for which function we use the simple notation $F(\xi)$. The straight line C has the equation

$$x = \xi + c(f(\xi))t = \xi + F(\xi)t , \tag{3.16}$$

where the function $F(\xi)$ is known, so that the curves C can be constructed in an (x, t) diagram. They are the *characteristic curves.* The solution to (3.10) satisfying the initial condition (3.15) can now be written

$$\rho = f(\xi) , \tag{3.17}$$

where ξ is implicitly given as a function of x and t through the equation (3.16). The proof that (3.17) is a solution of (3.10) is straight forward. The derivatives ρ_t and ρ_x are :

$$\rho_t = f'(\xi)\xi_t \tag{3.18}$$

$$\rho_x = f'(\xi)\xi_x , \tag{3.19}$$

where ξ_t and ξ_x are obtained by implicit derivation of (3.16):

$$0 = \xi_t + F'(\xi)\xi_t t + F(\xi) \tag{3.20}$$

$$1 = \xi_x + F'(\xi)\xi_x t . \tag{3.21}$$

Inserting ξ_t and ξ_x obtained from (3.20) and (3.21) into the expressions (3.18) and (3.19) for ρ_t and ρ_x we obtain

$$\rho_t = -\frac{F(\xi)f'(\xi)}{1+F'(\xi)t} \tag{3.22}$$

$$\rho_x = \frac{f'(\xi)}{1+F'(\xi)t} \, . \tag{3.23}$$

Because $F(\xi) = c(\rho)$ it follows immediately from (3.22) and (3.23) that (3.17) gives a solution to (3.10).

The straight lines C given by (3.16), in general, have different slopes. Therefore they must cross each other, which means that at some points in the (x,t) plane the ρ values are not unique. The manner through which the density $\rho(x,t)$ becomes multivalued in x for sufficiently large t values can be illustrated by a picture of the wave profile at two different times with a difference t.

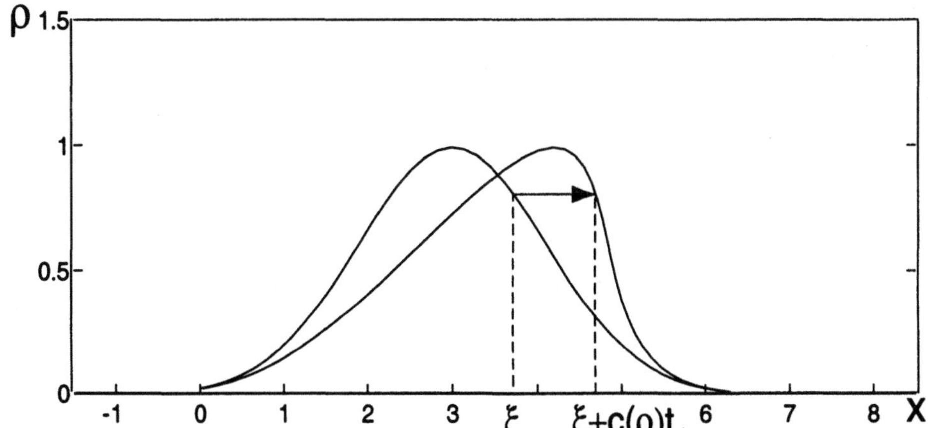

Figure 3.1: The wave profile $\rho(x)$ at the two times $t = 0$ and $t = t_1$.

3.1.3 Shock wave solution

It is easy to see from (3.13) and (3.14) that the wave profile at time t can be constructed from that at time $t = 0$ by moving each point on the profile a distance $c(\rho)t$ to the right. Thus for $c(\rho)$ an increasing function of ρ, which is usually the case, points with larger ρ move a longer distance to the right. This means that the wave profile becomes deformed and, as is seen from (3.22) and (3.23), ρ_x and ρ_t become infinite at

$$t = -\frac{1}{F'(\xi)} \ . \tag{3.24}$$

For t greater than the value in (3.24) the wave profile looks as in Figure 3.2, which means that, for some range of x values three ρ values correspond to the same value of x, e.g. around $x = 6$ in Figure 3.2.

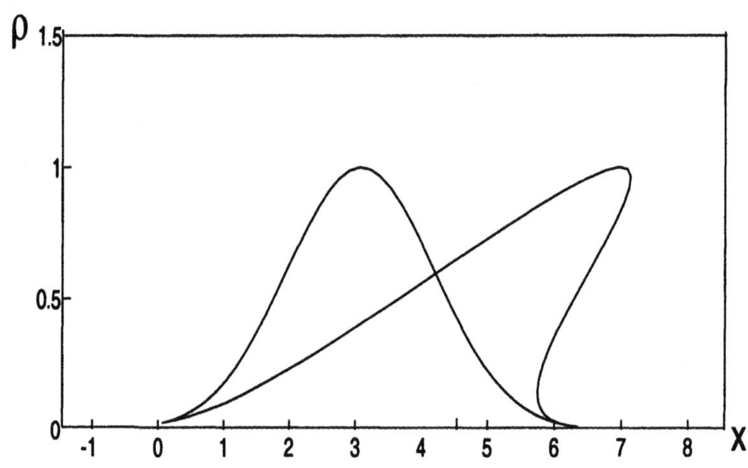

Figure 3.2: Wave profile for $t > -1/F'(\xi)$.

This is an impossible physical situation and therefore the mathematical model should be changed. However, if we allow discontinuities in the solution, we need not change the physical model (3.10). Instead we can replace the unacceptable part of a wave profile of the type shown in Figure 3.3. by a sudden change of ρ, a discontinuous shock.

In order to determine the velocity and position of the shocks, we assume that equation (3.10) describes a continuous mass distribution $\rho(x,t)$ moving with a velocity $v(x,t)$. The flux $q(x,t)$ is defined as

$$q = \rho v \ , \tag{3.25}$$

so that $q(x,t)$ is the mass per unit time passing the point x in a one-dimensional medium. We assume that the mass is conserved, which means that the change of the mass in an interval $x_1 < x < x_2$ must equal the net inflow or outflow, i.e.

$$\frac{d}{dt}\int_{x_1}^{x_2} \rho(x,t)dx = q(x_1,t) - q(x_2,t) \ . \tag{3.26}$$

With $x_2 \to x_1$ we obtain

$$\frac{\partial \rho}{\partial t} + \frac{\partial q}{\partial x} = 0 \ , \tag{3.27}$$

where the derivatives are assumed to be continuous. A functional connection between q and ρ is assumed,

$$q = Q(\rho) \ , \tag{3.28}$$

and so with

$$c(\rho) = Q'(\rho) \ , \tag{3.29}$$

the simple wave problem (3.10) is obtained. When the solution breaks down according to Figure 3.3 the derivability of ρ and q leading to (3.27) must be questioned as well as the relation (3.28) for discontinuous q and ρ. The mass conservation in the integral form (3.26) is, however, still valid. We now introduce a discontinuity in ρ, in other words - a shock, at $x = x_{sh}$ and assume that $x_{sh}(t)$ lies between x_1 and x_2. From (3.26) we then obtain

$$\begin{aligned} q(x_2,t) &- q(x_1,t) \\ &= \frac{d}{dt}\int_{x_1}^{x_{sh}(t)} \rho(x,t)dx + \frac{d}{dt}\int_{x_{sh}(t)}^{x_2} \rho(x,t)\, dt \\ &= \rho(x_{sh}^-,t)\dot{x}_{sh} - \rho(x_{sh}^+,t)\dot{x}_{sh} + \int_{x_1}^{x_{sh}(t)} \rho_t(x,t)dx + \int_{x_{sh}(t)}^{x_2} \rho_t(x,t)dx \ . \end{aligned} \tag{3.30}$$

If $x_1 \to x_2$ we obtain, because ρ_t is bounded in both intervals in (3.30)

$$q(x_{sh}^-,t) - q(x_{sh}^+,t) = [\rho(x_{sh}^-,t) - \rho(x_{sh}^+,t)]\dot{x}_{sh} \ . \tag{3.31}$$

If $v_{sh} \equiv \dot{x}_{sh}$ is the velocity of the discontinuity and values of variables to the left of the shock are indicated by '1' and to the right of the shock by '2', from (3.31) we obtain the Rankine-Hugoniot relation (Rankine 1870; Hugoniot 1887, 1889)

$$q_2 - q_1 = v_{sh}(\rho_2 - \rho_1) \ . \tag{3.32}$$

The conservation of momentum and energy leads to two more Rankine-Hugoniot relations (Pierce 1981). Thus in the limit

$$\frac{\rho_2 - \rho_1}{\rho_1} \to 0 \tag{3.33}$$

the shock velocity v_{sh} approaches the characteristic velocity $c(\rho)$ given by (3.29). A Taylor expansion of v_{sh} using (3.32) and (3.28) gives

$$v_{sh} = \frac{1}{\rho_2 - \rho_1}[(\rho_2 - \rho_1)Q'(\rho_1) + \frac{1}{2}(\rho_2 - \rho_1)^2 Q''(\rho_1) + ...] \ . \tag{3.34}$$

From (3.29) we also have the Taylor expansion

$$c(\rho_2) = c(\rho_1) + (\rho_2 - \rho_1)Q''(\rho_1) \ . \tag{3.35}$$

Comparison between (3.34) and (3.35) gives

$$v_{sh} = \frac{1}{2}(c_1 + c_2) + O((\rho_2 - \rho_1)^2) \ . \tag{3.36}$$

In the so called weak shock approximation

$$\frac{\rho_2 - \rho_1}{\rho_1} \ll 1 \ , \tag{3.37}$$

thus the shock velocity is the average of the characteristic velocities on both sides of the shock.

3.1.4 Rule of equal areas

We will now find a simple way to construct the position x_{sh} of the shock originating from the multivaluedness of the solution (3.17) with

(3.16) of the equation (3.10). For a sufficiently large value of t, this solution becomes multivalued in the interval $x_a < x < x_b$, as shown in Figure 3.3. The function ρ is made unique by imposing a discontinuity at $x = x_{sh}$.

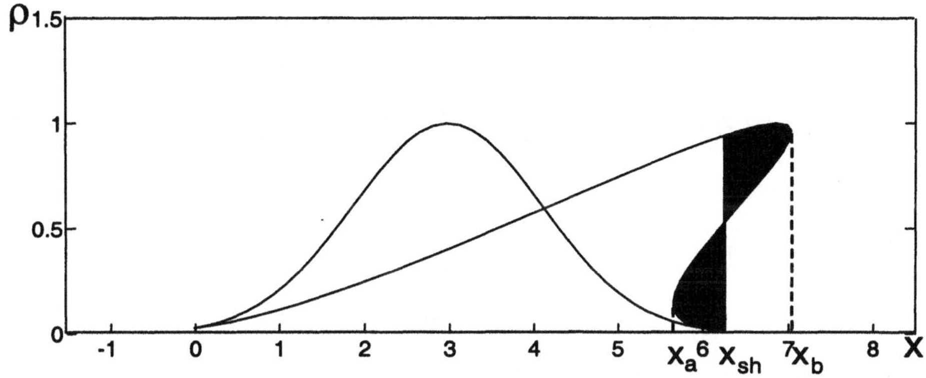

Figure 3.3. Replacement of the multivalued part of $\rho(x)$ in Figure 3.2 by a discontinuity.

We will show *the rule of equal areas* which means that the position of x_{sh} makes the shadowed areas in Figure 3.3 equal. The shadowed area $A(t)$ is with its sign

$$A(t) = -\int (x - x_{sh})d\rho = -\int_{\xi_-(t)}^{\xi_+(t)} [x(\xi,t) - x_{sh}(t)] \frac{df(\xi)}{d\xi}\, d\xi \;, \quad (3.38)$$

where ξ_+ and ξ_- are the two ξ values giving the shock position according to (3.16):

$$x_{sh} = \xi_- + F(\xi_-)t = \xi_+ + F(\xi_+)t \;. \quad (3.39)$$

The equation (3.39) expresses the fact that, although $f(\xi)$ is a one-valued function, the function $\rho(x,t) = f(\xi(x,t))$ is a multi-valued function of x for a given t value since the x value x_{sh} corresponds to both ξ_+ and ξ_- with different values of $f(\xi)$. We now evaluate the derivative dA/dt from (3.38), using (3.16):

$$\begin{aligned}\frac{dA}{dt} &= -\int_{\xi_-(t)}^{\xi_+(t)} \left(\frac{\partial x(\xi,t)}{\partial t} - \frac{dx_{sh}(t)}{dt}\right)\frac{df(\xi)}{d\xi}\, d\xi \\ &= -\int_{\xi_-(t)}^{\xi_+(t)} \left(F(\xi) - \frac{dx_{sh}(t)}{dt}\right)\frac{df(\xi)}{d\xi}\, d\xi \;. \quad (3.40)\end{aligned}$$

According to (3.29)

$$F(\xi) = c(f(\xi)) = Q'(f(\xi)) \tag{3.41}$$

we obtain by integration of (3.40) the equation

$$\frac{dA}{dt} = -Q(f(\xi_+)) + Q(f(\xi_-)) + \frac{dx_{sh}(t)}{dt}(f(\xi_+) - f(\xi_-)) \; . \tag{3.42}$$

Since

$$v_{sh} = \frac{dx_{sh}(t)}{dt} \tag{3.43}$$

a comparison of (3.42) and (3.32) gives

$$\frac{dA}{dt} = 0 \; . \tag{3.44}$$

However, $A(t) = 0$ at the moment when the discontinuity first appears. Thus $A(t) = 0$ for all t and the rule of equal areas, visualized in Figure 3.3 , is verified.

If (3.10) is put into another form, then the rule of equal areas can be given a form easier to use than that shown in Figure 3.3. Multiplying (3.10) by $c'(\rho)$ we find

$$c_t + cc_x = 0 \; . \tag{3.45}$$

This equation is essentially identical with Burgers' equation (2.79) if we neglect the second derivative term. Then we obtain the Riemann equation, for the initial value problem written as

$$\frac{\partial v}{\partial t} + \beta v \frac{\partial v}{\partial \xi} - = 0 \; . \tag{3.46}$$

We assume that c is a decreasing function of ρ, i.e. $c'(\rho) < 0$. The picture corresponding to Figure 3.3 is then as shown in Figure 3.4.

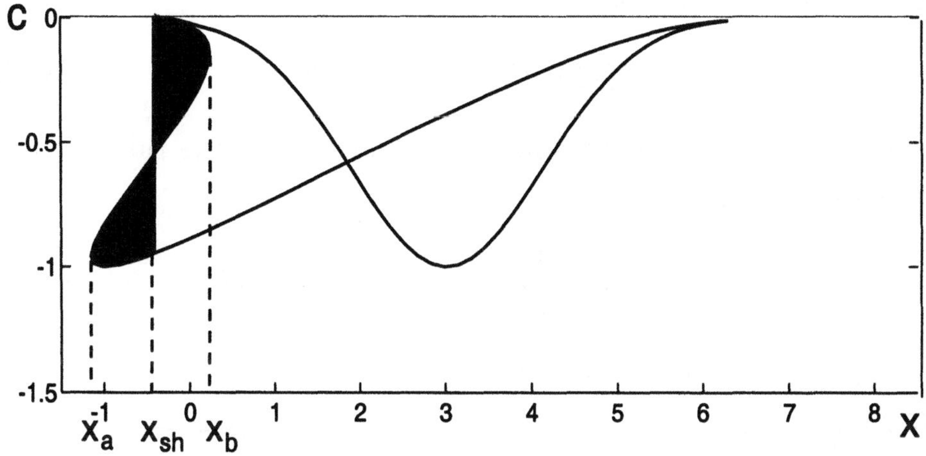

Figure 3.4 Replacing the multivalued part of $c(x)$ by a discontinuity.

The rule of equal areas (3.38) with $A(t) = 0$ is valid also for the solution

$$c = F(\xi) \tag{3.47}$$

of (3.45), with ξ given by (3.16). Thus we obtain from (3.38) and (3.47)

$$0 = -\int_{\xi_-(t)}^{\xi_+(t)} (x(\xi,t) - x_{sh}(t)) \frac{dF(\xi)}{d\xi}\, d\xi \;. \tag{3.48}$$

A partial integration of (3.48) gives

$$0 = -\int_{\xi_-(t)}^{\xi_+(t)} \frac{\partial x(\xi,t)}{\partial \xi} F(\xi)\, d\xi \;, \tag{3.49}$$

since the parenthesis of the integrand vanishes in the limits .

Using (3.16) we obtain from (3.49)

$$0 = -\int_{\xi_-(t)}^{\xi_+(t)} (1 + F'(\xi)t) F(\xi) d\xi \;, \tag{3.50}$$

that is,

$$\int_{\xi_-(t)}^{\xi_+(t)} F(\xi) d\xi = -\frac{1}{2} t [F^2(\xi_+) - F^2(\xi_-)] \;. \tag{3.51}$$

Since, according to (3.39)

$$[F(\xi_+) - F(\xi_-)]t = \xi_- - \xi_+ , \tag{3.52}$$

we obtain from (3.51) the following formulation of the rule of equal areas in the (c, ξ) plane

$$\int_{\xi_-(t)}^{\xi_+(t)} F(\xi)d\xi = \frac{1}{2}(\xi_+ - \xi_-)[F(\xi_+) + F(\xi_-)] . \tag{3.53}$$

The equation (3.53) is represented geometrically in Figure 3.5.

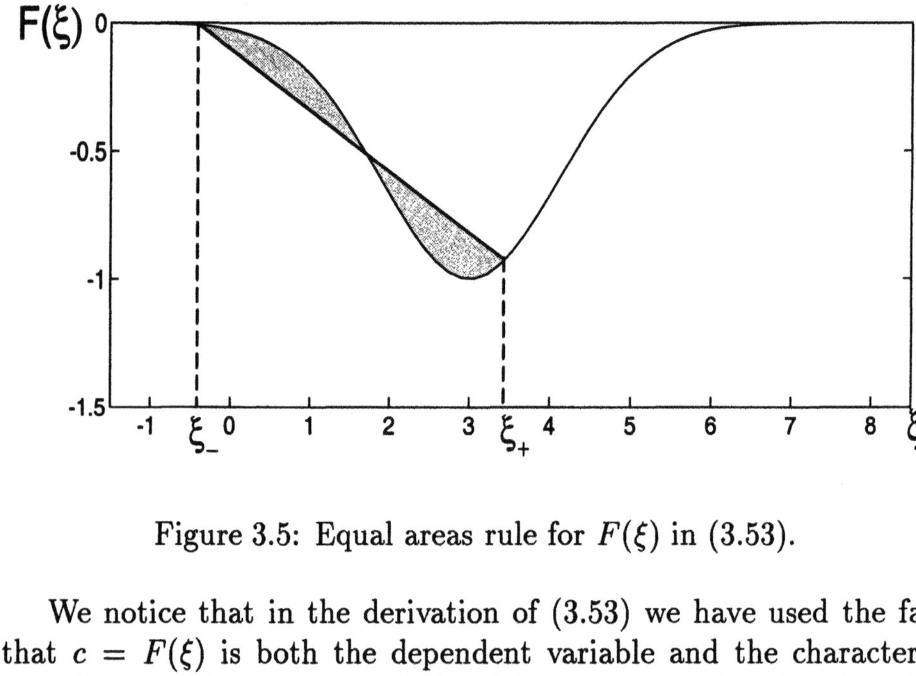

Figure 3.5: Equal areas rule for $F(\xi)$ in (3.53).

We notice that in the derivation of (3.53) we have used the fact that $c = F(\xi)$ is both the dependent variable and the characteristic velocity. This property of the differential equation (3.45) makes the integration of (3.50) possible. Investigations of Rott (1978, 1980) present corrections to x_{sh} and v_{sh} using Lagrangian coordinates.

3.1.5 Prediction of wave behaviour from area differences

In the following will be described a time domain solution useful for general boundary conditions, which are not strictly positive or strictly negative pulses. A qualitative prediction of the longterm behaviour can be made from analyzing the area differences between positive and negative parts of the boundary condition.

The choice of form for the Riemann equation is the dimensionless equation (cf. (2.76) with $\epsilon = 0$):

$$\frac{\partial V}{\partial \sigma} - V\frac{\partial V}{\partial \theta} = 0 \ . \tag{3.54}$$

For the boundary conditon

$$V(\sigma = 0, \theta) = V_0(\theta) \tag{3.55}$$

a solution is

$$V(\sigma, \theta) = V_0(\theta_m(\sigma, \theta)) \ , \tag{3.56}$$

where $\theta_m(\sigma, \theta)$ is defined implicitly as the solution θ_0, of the equation

$$\theta = \theta_0 - \sigma V_0(\theta_0) \ . \tag{3.57}$$

Another version of the solution is given from (3.56)-(3.57)

$$V(\sigma, \theta) = \frac{\theta_m - \theta}{\sigma} \ , \tag{3.58}$$

with θ_m again given by (3.57).

There exists a general way of predicting the qualitative evolution for the Riemann equation by looking at the area under the positive and negative parts of the boundary condition (Novikov 1978). The main characteristics of wave behaviour can be seen by comparing the areas (integrals) of the boundary condition (3.55) between the zeros with negative derivative (the non-shock zeros). These area differences will be denoted ΔA. The area difference $\Delta A(\sigma)$ is constant, $\Delta A(\sigma) = \Delta A_0$, for Riemann waves. This will now be shown by starting with the Riemann equation (3.54), written as

$$V_\sigma - VV_\theta = 0 \tag{3.59}$$

Let θ_n, $n = 1, 2, \ldots$, be the positions where $V = 0$ and integrate between two zeros :

$$\int_{\theta_n}^{\theta_{n+1}} (V_\sigma - VV_\theta)d\theta = 0 \ . \tag{3.60}$$

Assuming that we do not have any shocks moving in from the outside:

$$\int_{\theta_n}^{\theta_{n+1}} V_\sigma d\theta = \frac{d}{d\sigma}\int_{\theta_n}^{\theta_{n+1}} V d\theta = \frac{d}{d\sigma}\Delta A(\sigma, n) \tag{3.61}$$

$$\int_{\theta_n}^{\theta_{n+1}} V V_\theta \, d\theta = [VV]_{\theta_n}^{\theta_{n+1}} = 0 \; . \tag{3.62}$$

The equations (3.60) - (3.62) give

$$\frac{d}{d\sigma}\Delta A(\sigma, n) = 0 \rightarrow \Delta A(\sigma, n) = \Delta A_0(n) = constant \tag{3.63}$$

Because these area differences are constants it is simplest to obtain them from the boundary condition. An graphic example is shown in Figure 3.7. If positioned at the zeros with positive derivative (the shock zeros) the $\Delta A_0(n)$s are describing the velocity of shocks and the future wave form can be estimated (Gurbatov and Hedberg 1998). Positive $A_0(n)$ will be moving to the left and negative to the right. Larger amplitudes will move faster and when colliding the amplitude values are added together. For example, one positive and one negative shock with equal amplitudes ΔA_0 will cancel each other when colliding and the remaining shock will stand still.

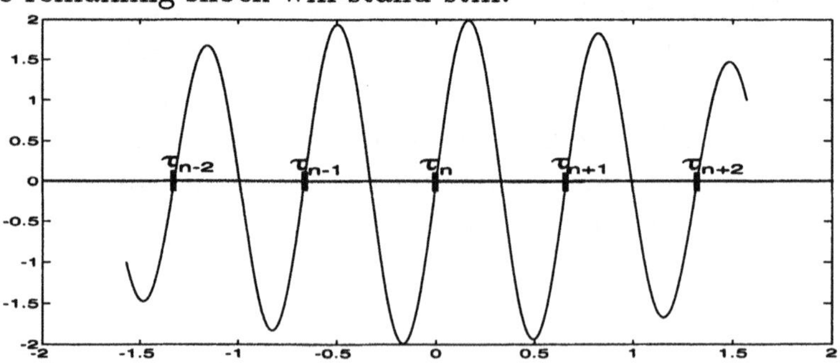

Figure 3.6: The integrations, in order to obtain the area differences ΔA_0, are made between the θ_ns, which are the zeroes of $V_0(\theta)$ in the boundary condition (3.55).

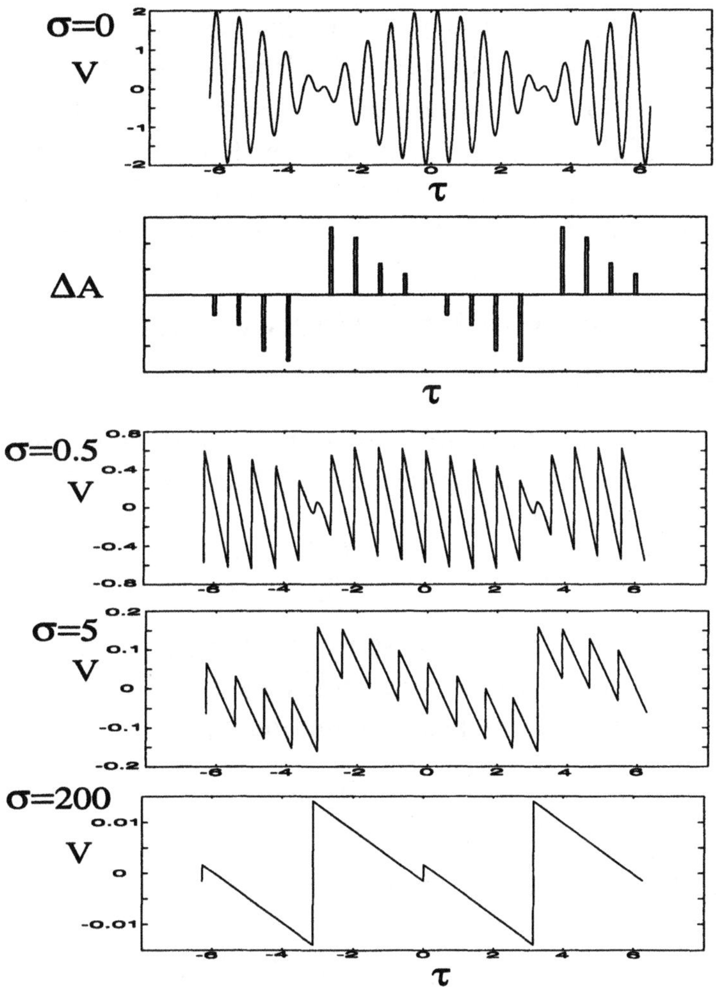

Figure 3.7: An example of how the area differences ΔA_0 help predict wave development; $V_0 = \sin 9\theta + \sin 10\theta$.

Note that these graphs do not predict the amplitude, only the velocity, of shocks. Thus, shocks which are standing still will not be seen in the ΔA_0-plots .

3.2 Exact solution of Burgers' equation

Although the Riemann equation treated in the preceeding section is in many cases a useful model of nonlinear acoustic wave propagation, a better model is needed when the dissipative effects on the wave propagation are of interest. This model is Burgers' equation, for which an exact solution for plane waves will be given in this section. The discontinuities of the Riemann equation are then avoided. In the end of this section, Burgers' equation will be used in the limit of vanishing viscocity to obtain alternative formulations of zero diffusion waves.

3.2.1 The Cole-Hopf solution of Burgers' equation

The nonlinear equation for plane waves in a homogeneous dissipative medium is Burgers' equation, given in (2.76) and rewritten here:

$$\frac{\partial V}{\partial \sigma} - V\frac{\partial V}{\partial \theta} - \epsilon\frac{\partial^2 V}{\partial \theta^2} = 0 \ . \tag{3.64}$$

The equation (3.64) can be transformed into a linear equation. The transformation is easiest to see through if we start with the equivalent linear equation, which is the heat conduction equation

$$\frac{\partial U}{\partial \sigma} = \epsilon\frac{\partial^2 U}{\partial \theta^2} \ . \tag{3.65}$$

In equation (3.65) we try a solution of the form

$$U = C\exp\Big(\frac{1}{2\epsilon}\int_0^\theta V(\sigma,\theta')\,d\theta'\Big) \ . \tag{3.66}$$

Derivation of (3.66) gives

$$\frac{\partial U}{\partial \sigma} = \frac{C}{2\epsilon}\int_0^\theta \frac{\partial V}{\partial \sigma}\,d\theta' \exp\Big(\frac{1}{2\epsilon}\int_0^\theta V\,d\theta'\Big) \tag{3.67}$$

$$\frac{\partial U}{\partial \theta} = \frac{C}{2\epsilon}V\exp\Big(\frac{1}{2\epsilon}\int_0^\theta V\,d\theta'\Big) \tag{3.68}$$

$$\frac{\partial^2 U}{\partial \theta^2} = \Big(\frac{C}{2\epsilon}\frac{\partial V}{\partial \theta} + \frac{C}{4\epsilon^2}V^2\Big)\exp\Big(\frac{1}{2\epsilon}\int_0^\theta V\,d\theta'\Big) \ . \tag{3.69}$$

Insertion of (3.67) and (3.69) into (3.65) gives

$$\int_0^\theta \frac{\partial V}{\partial \sigma}\,d\theta' = \epsilon\frac{\partial V}{\partial \theta} + \frac{1}{2}V^2 \ . \tag{3.70}$$

Derivation of (3.70) with respect to θ gives Burgers' equation (2.76). The transformation (3.66) thus aims to rewrite the two terms with θ-derivatives in (2.76) as a single second θ-derivative. The transformation which linearizes the nonlinear equation (2.76), is obtained from (3.66)

$$V = 2\epsilon \frac{\frac{\partial U}{\partial \theta}}{U} = 2\epsilon \frac{\partial}{\partial \theta}(\ln U) \ . \tag{3.71}$$

This transformation was found by E. Hopf (1950) and J. D. Cole (1951). In order to find the solution of Burgers' equation (2.76) with a given boundary condition on V for $\sigma = 0$, this boundary condition is transformed into a boundary condition on U by means of (3.66). Then the heat conduction equation (3.65) is solved with the boundary condition

$$U(\sigma = 0, \theta) = U_0(\theta) \ . \tag{3.72}$$

The solution of the heat conduction equation (3.65) can be found by separation of the variables:

$$U(\sigma, \theta) = S(\sigma)T(\theta) \ . \tag{3.73}$$

Insertion of (3.73) into (3.65) gives

$$\frac{dS}{d\sigma}\frac{1}{S} = \epsilon \frac{d^2T}{d\theta^2}\frac{1}{T} \ . \tag{3.74}$$

Since the members of (3.74) depend only on σ and θ respectively, they must be equal to the same constant, which we call $-\epsilon k^2$. Each k value thus gives a solution

$$U_k(\sigma, \theta) = c_k S_k(\sigma) T_k(\theta) \ . \tag{3.75}$$

where c_k is a constant and

$$S_k = \mathrm{e}^{-\epsilon k^2 \sigma} \tag{3.76}$$

$$T_k = \mathrm{e}^{ik\theta} \ . \tag{3.77}$$

The general solution of (3.65) can be written as a sum

$$U = \sum_{k=-\infty}^{\infty} c_k S_k T_k \ , \tag{3.78}$$

or as an integral

$$U(\sigma,\theta) = \int_{-\infty}^{\infty} c(k)\mathrm{e}^{ik\theta-\epsilon k^2\sigma}\,dk\ . \tag{3.79}$$

In the integral form (3.79) we determine the function $c(k)$ by means of the boundary condition (3.72). Inserting (3.72) into (3.78) and using the inversion of the Fourier transform we find

$$c(k) = \frac{1}{2\pi}\int_{-\infty}^{\infty} U_0(\theta')\mathrm{e}^{-ik\theta'}\,d\theta'\ . \tag{3.80}$$

Inserting (3.80) into (3.79) and evaluating the k integral we obtain the solution

$$U(\sigma,\theta) = \frac{1}{2\sqrt{\pi\epsilon\sigma}}\int_{-\infty}^{\infty} U_0(\theta')\mathrm{e}^{-\frac{(\theta-\theta')^2}{4\epsilon\sigma}}\,d\theta'\ . \tag{3.81}$$

Given a boundary condition

$$V(\sigma=0,\theta) = V_0(\theta) \tag{3.82}$$

of Burgers' equation (2.76), U_0 to be inserted in (3.81) is obtained from (3.66).

Another way of representing the solution of (3.65) is as a Fourier series. This representation is appropriate when the boundary condition is periodical. The solution (3.78) is written as

$$U(\sigma,\theta) = \sum_{k=0}^{\infty} A_k\mathrm{e}^{-\epsilon k^2\sigma}\cos k\theta\ . \tag{3.83}$$

In addition to the boundary condition (3.82) we now require periodicity for $\sigma > 0$:

$$V(\sigma,\theta=0) = V(\sigma,\theta=\pi) = 0\ . \tag{3.84}$$

The conditions (3.82) and (3.84) through (3.66) and (3.68) give the following conditions on $U(\sigma,\theta)$:

$$U(\sigma=0,\theta) = U_0(\theta) = C\exp\left(\frac{1}{2\epsilon}\int_0^{\theta} V_0(\theta')\,d\theta'\right)\ . \tag{3.85}$$

$$U_\theta(\sigma,\theta=0) = U_\theta(\sigma,\theta=\pi) = 0 \tag{3.86}$$

The condition (3.86) is satisfied by the Fourier expansion (3.83). The coefficients A_k in (3.83) can be determined using (3.85):

$$A_0 = \frac{1}{\pi}\int_0^\pi U_0(\theta)\, d\theta = \frac{C}{\pi}\int_0^\pi \exp\Big(\frac{1}{2\epsilon}\int_0^\theta V_0(\theta')\, d\theta'\Big)\, d\theta \qquad (3.87)$$

$$\begin{aligned} A_k &= \frac{2}{\pi}\int_0^\pi U_0(\theta)\cos k\theta\, d\theta \\ &= \frac{2C}{\pi}\int_0^\pi \exp\Big(\frac{1}{2\epsilon}\int_0^\theta V_0(\theta')d\theta'\Big)\cos k\theta\, d\theta, \quad k = 1, 2, 3 \ldots \end{aligned} \qquad (3.88)$$

Using either (3.81) or (3.83) with (3.87) and (3.88) we now know the exact solution of Burgers' equation for a given boundary condition using (3.71). However, the solutions are, although exact, in an awkward form.

The solution to Burgers' equation (3.64) through the Cole-Hopf transformation (3.71) is – with (3.65), (3.81) and (3.85):

$$\begin{aligned} V &= 2\epsilon\frac{\partial}{\partial\theta}[\ln(\frac{C}{2\sqrt{\pi\epsilon\sigma}}\int_{-\infty}^{\infty}\exp(F(\theta'))\, \mathrm{d}\theta')] \\ &= 2\epsilon[\int_{-\infty}^{\infty}\frac{\partial F}{\partial\theta}\mathrm{e}^{F(\theta')}\, \mathrm{d}\theta']/[\int_{-\infty}^{\infty}\mathrm{e}^{F(\theta')}\, \mathrm{d}\theta'] \\ &= [\int_{-\infty}^{\infty}\frac{\theta'-\theta}{\sigma}\mathrm{e}^{F(\theta')}\, \mathrm{d}\theta']/[\int_{-\infty}^{\infty}\mathrm{e}^{F(\theta')}\, \mathrm{d}\theta']\,, \end{aligned} \qquad (3.89)$$

where

$$F(\theta') = \frac{1}{2\epsilon}[\int^{\theta'} V_0(\theta'')\, d\theta'' - \frac{(\theta'-\theta)^2}{2\sigma}]\,. \qquad (3.90)$$

In our forthcoming treatment of the evolution of different wave forms given at the boundary we will attempt to obtain simpler approximate solutions from (3.71).

3.2.2 Burgers' equation with vanishing diffusivity

Look at the form of the solution (3.89) with (3.90) of Burgers' equation (3.64). As the exponent varies greatly the main contribution to the integral is from around the maxima of the function f. Often only one maximum is necessary at any specific point in time θ. For this case the saddle-point approach gives very good results through approximating the function around the maxima.

In the case of vanishing diffusion, letting $\epsilon \to 0$, by the saddle-point method (or steepest descent) (Bender and Orszag 1978) it follows that the total contribution to the integrals comes from the point θ_0 which is situated at the maximum of $F(\theta')$: $Max\{F(\theta')\} = F(\theta_0)$. One saddle-point is then at any specific distance σ and time θ describing the solution. The fraction inside the top integral (3.89) may be put in front of the integrals, which then cancel each other, so that

$$V(\theta,\sigma) = \frac{\theta_0 - \theta}{\sigma} . \tag{3.91}$$

Because the maximum of $F(\theta')$, equalling $F(\theta_0)$, is given by

$$\frac{dF(\theta')}{d\theta'} = 0 \tag{3.92}$$

we obtain with (3.90)

$$\frac{\partial F}{\partial \theta'} = \frac{1}{2\epsilon}[V_0 - \frac{\theta' - \theta}{\sigma}] = 0. \tag{3.93}$$

This results in

$$V_0(\theta_0) = \frac{\theta_0 - \theta}{\sigma} = V(\theta,\sigma) , \tag{3.94}$$

where the last equality comes from (3.91). This is the solution of Burgers' equation for vanishing ϵ.

This can now be compared to the solution of the corresponding equation for $\epsilon = 0$, the Riemann equation, here written in the same dimensionless variables as the Burgers equation (2.76):

$$\frac{\partial V}{\partial \sigma} - V\frac{\partial V}{\partial \theta} = 0 . \tag{3.95}$$

The straightforward approach is to insert the solution (3.94) into (3.95) to see if this will solve the equation exactly. Using (3.94) we make the derivations

$$\frac{\partial V}{\partial \sigma} = \frac{\partial V_0}{\partial \theta_0}\frac{\partial \theta_0}{\partial \sigma} \tag{3.96}$$

$$\frac{\partial V}{\partial \theta} = \frac{\partial V_0}{\partial \theta_0}\frac{\partial \theta_0}{\partial \theta} . \tag{3.97}$$

From the last part of (3.94), written as

$$\theta_0 = \theta + \sigma V(\theta, \sigma) \tag{3.98}$$

we now find the derivatives

$$\frac{\partial \theta_0}{\partial \sigma} = V(\sigma, \theta) + \sigma \frac{\partial V}{\partial \sigma} \tag{3.99}$$

$$\frac{\partial \theta_0}{\partial \theta} = 1 + \sigma \frac{\partial V}{\partial \theta} . \tag{3.100}$$

Inserting (3.99) into (3.96) and (3.100) into (3.97) we evaluate the derivatives $\frac{\partial V}{\partial \sigma}$ and $\frac{\partial V}{\partial \theta}$ in terms of the function $\frac{\partial V_0}{\partial \theta_0}$, which is known by the boundary condition (3.82):

$$\frac{\partial V}{\partial \sigma} = V \frac{\partial V_0}{\partial \theta_0} \frac{1}{1 - \sigma \frac{\partial V_0}{\partial \theta_0}} \tag{3.101}$$

$$\frac{\partial V}{\partial \theta} = \frac{\partial V_0}{\partial \theta_0} \frac{1}{1 - \sigma \frac{\partial V_0}{\partial \theta_0}} . \tag{3.102}$$

From (3.101) and (3.102) we see that (3.95) is exactly solved by (3.94). The case of vanishing diffusion is equal to that of zero diffusion.

Chapter 4

Nonlinear waves with zero and vanishing diffusion

The method of solving Riemann's equation was developed in the preceding chapter for arbitrary original wave profiles. The method resulted in general rules for predicting the evolution of shock wave profiles. In this chapter these rules will be applied to specific profiles namely short pulses, sinusoidal waves, and multifrequency waves.

4.1 Short pulses

4.1.1 Triangular pulses

The evolution of a short pulse according to the Riemann wave equation (3.45) will now be studied, here rewritten in (4.1),

$$c_t + cc_x = 0 \ , \tag{4.1}$$

using the rule of equal areas in the form (3.53). We assume that the initial pulse is

$$c(x, 0) = F(x) \ , \tag{4.2}$$

where $F(x) = c_0$ for $x < -l$ and for $x > 0$.

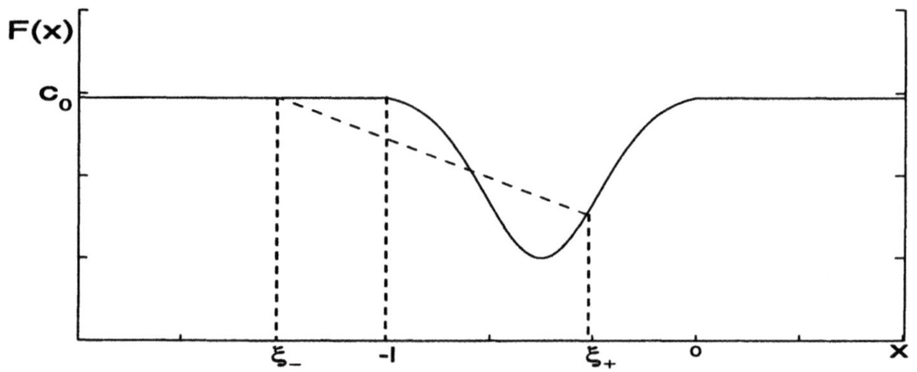

Figure 4.1: Equal areas rule for $F(x)$ in (4.3).

Since $\xi_- < -l$ and $F(\xi_-) = c_0$ (see Figure 4.1) the rule of equal areas (3.53) can be written

$$\frac{1}{2}(c_0 + F(\xi_+))(\xi_+ - \xi_-) = \int_{\xi_-(t)}^{\xi_+(t)} F(\xi)d\xi \, , \tag{4.3}$$

Combining (4.3) with the identity

$$c_0(\xi_+ - \xi_-) = \int_{\xi_-(t)}^{\xi_+(t)} c_0 \, d\xi \tag{4.4}$$

we obtain

$$\frac{1}{2}(c_0 - F(\xi_+))(\xi_+ - \xi_-) = \int_{-l}^{\xi_+(t)} (c_0 - F(\xi))d\xi \, , \tag{4.5}$$

where we have replaced the lower limit ξ_- of the righthand side integral by $-l$. This can be done since $\xi_- < -l$ and $F(\xi) = c_0$ for $\xi < -l$. Using the relation

$$t = \frac{\xi_+ - \xi_-}{c_0 - F(\xi_+)} \tag{4.6}$$

obtained from (3.39), we rewrite (4.5):

$$\frac{1}{2}(c_0 - F(\xi_+))^2 t = \int_{-l}^{\xi_+(t)} (c_0 - F(\xi))d\xi \, . \tag{4.7}$$

The position of the discontinuity $x_{sh}(t)$ and the value c_+ of $c(x,t)$ just to the right of the discontinuity are given by

$$x_{sh}(t) = \xi_+ + F(\xi_+)t \tag{4.8}$$

and

$$c_+ = F(\xi_+) \ , \tag{4.9}$$

where $\xi_+(t)$ is found from (4.7). When t approaches infinity ξ_+ approaches zero and (4.7) gives

$$F(\xi_+) \approx c_0 - \sqrt{\frac{2A}{t}} \ , \tag{4.10}$$

where A is the area of depression below the value c_0 of $F(\xi)$:

$$A = \int_{-l}^{0} (c_0 - F(\xi))d\xi \ . \tag{4.11}$$

Using (4.10) and the fact that $F(\xi_+) \to c_0$ we find the asymptotic forms of the equations (4.8) and (4.9) for t approaching infinity:

$$x_{sh}(t) \approx c_0 t - \sqrt{2At} \tag{4.12}$$

$$c_+ - c_0 = -\sqrt{\frac{2A}{t}} \ . \tag{4.13}$$

The solution $c(x,t)$ to the right of the discontinuity is given by (3.47) and (3.16), where

$$\xi_+ < \xi < 0 \ . \tag{4.14}$$

Since $\xi_+ \to 0$ for $t \to \infty$, those values of ξ satisfying the inequality (4.14) also approach zero; therefore from (3.16) and (3.47) we obtain

$$c \approx \frac{x}{t} \qquad \text{as } t \to \infty \ . \tag{4.15}$$

Using (4.15) and the fact that

$$c_+ < c < c_0 \tag{4.16}$$

for the depression of the solution $c(x,t)$ to the right of the discontinuity, we obtain from (4.13) the result

$$c_0 t - \sqrt{2At} < x < c_0 t \ . \tag{4.17}$$

The asymptotic solution is shown in Figure 4.2.

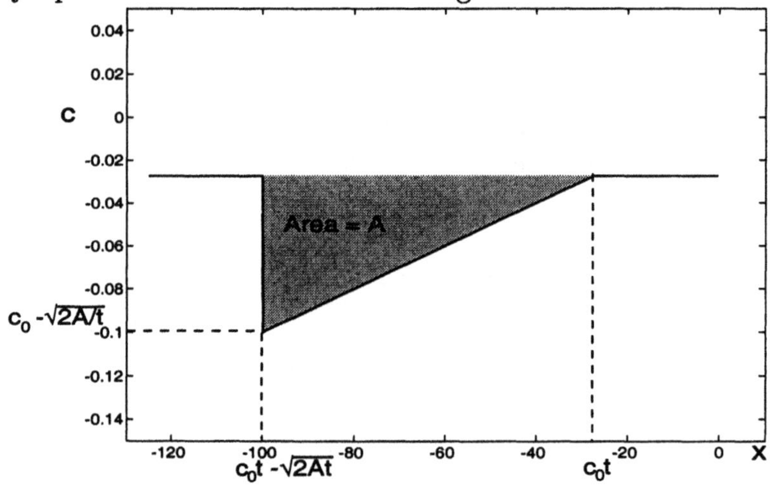

Figure 4.2: Asymptotic solution of (3.45) with the initial pulse given by Figure 4.1.

It is seen that all structure in the initial pulse (4.2) has disappeared in the asymptotic pulse.

A theoretical and experimental study of the nonlinear lengthening of a triangular acoustic pulse was made by Temkin and Maxham (1985). The evolution of simple short pulses is more extensively accounted for by Whitham (1974).

4.1.2 N-waves

The treatment of short triangular pulses just completed can easily be extended to the so called *N-waves*, named for their similarity with the letter N. They will be studied more extensively in chapter 5 and 6. However, preparatory to that we make the following changes of variables in (3.45):

$$C = c - c_0 \tag{4.18}$$

$$y = x - c_0 t \ . \tag{4.19}$$

We then obtain

$$C_t + CC_y = 0 \ . \tag{4.20}$$

Equation (4.20) is invariant for the change $y \to -y$, $C \to -C$ and is therefore solved by the N-wave shown in Figure 4.3. The next step is to make the variables dimensionless by scaling. The dimensionless variables are:

$$T = \frac{t}{t_0} \tag{4.21}$$

$$X = \frac{y}{y_0} \tag{4.22}$$

$$V = \frac{C\,t_0}{y_0} \ , \tag{4.23}$$

where t_0 is the time at which the triangular pulse in Figure 4.3. is fully developed from the initial disturbance (4.2) and y_0 is given by

$$y_0 = \sqrt{2At_0} \ , \tag{4.24}$$

where A is the area under the triangular pulse. The dimensionless equation

$$\frac{\partial V}{\partial T} + V\frac{\partial V}{\partial X} = 0 \tag{4.25}$$

has the N-wave solution $V = V_0(X,T)$ given as

$$V_0(X,T) = \frac{X}{T} \ , \qquad |X| < T^{\frac{1}{2}} \tag{4.26}$$

$$V_0(X,T) = 0 \ , \qquad |X| > T^{\frac{1}{2}} \tag{4.27}$$

satisfying the initial condition

$$V_0(X,1) = X \ , \qquad |X| < 1 \tag{4.28}$$

$$V_0(X,1) = 0 \ , \qquad |X| > 1 \ . \tag{4.29}$$

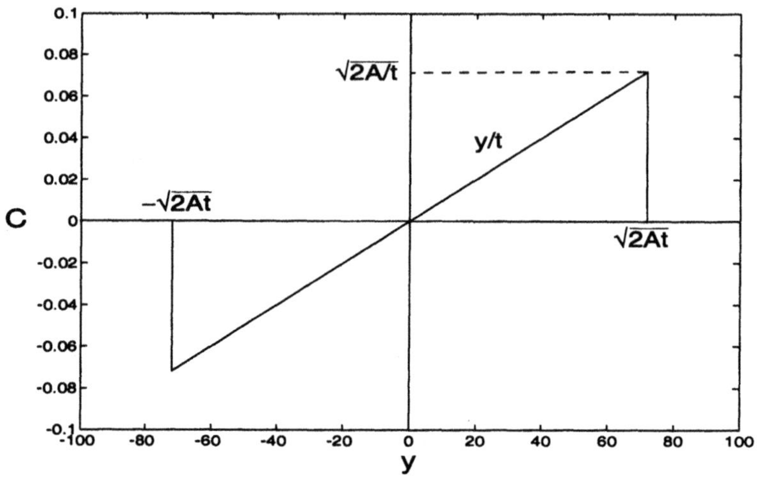

Figure 4.3: The N-wave solution of (4.20).

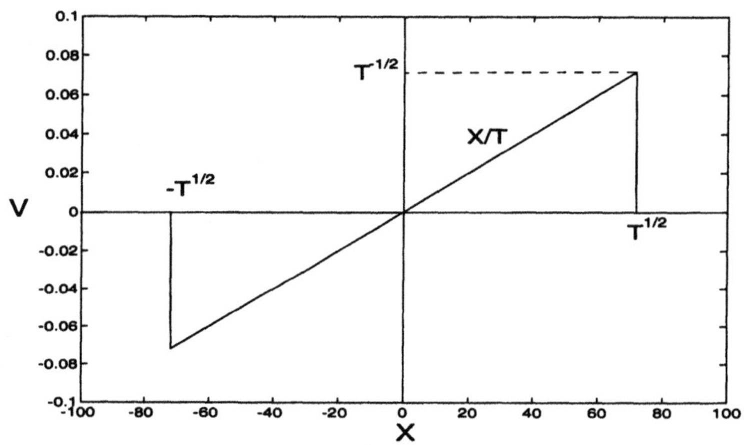

Figure 4.4: The N-wave solution of (4.25).

The solution (4.26) and (4.27) in $X < 0$ is, due to the transformations (4.18) and (4.19) and (4.21) -(4.24), identical with the triangular pulse solution (4.15) between the limits (4.17) shown in Figure 4.2.

The study of the evolution of the initial short pulses developing into triangular waves or N-waves made by using the Riemann wave equation (3.45) (or (2.79) neglecting the second derivative term) can equally well be done for Burgers' equation (2.71), neglecting the second derivative term, if the initial condition is replaced by a boundary condition. The N-wave is the final profile of an arbitrary finite pulse with zero mean value, as is shown by Gurbatov, Enflo and Pasmanik (1999).

4.2 Sinusoidal waves

4.2.1 Continuous solution

The solution (3.16) and (3.17) of the Riemann wave (3.10) will now be used for studying the evolution of a strong plane sine wave generated at the end of a long pipe filled with a homogeneous lossless fluid. The latter assumption implies that our model equation is (2.71) without the second derivative term

$$\frac{\partial v}{\partial x} - \frac{\beta}{c_0^2} v \frac{\partial v}{\partial \tau} = 0 \ , \tag{4.30}$$

where $\tau = t - x/c_0$.

Since $\tau = t$ for $x = 0$ the boundary condition

$$v(x = 0, t) = v_0 \sin \omega t \tag{4.31}$$

can be formulated as

$$v(x = 0, \tau) = v_0 \sin \omega \tau \ . \tag{4.32}$$

As (4.30) is a special case of (3.10) with the independent variables (x, t) replaced by (x, τ), the solution of equation (4.30) with the boundary condition (4.32) can be written down in analogy with the solution (3.16) and (3.17):

$$v = v_0 \sin \omega \Phi \ , \tag{4.33}$$

with Φ given by

$$\Phi = \tau + \frac{\beta v}{c_0^2} x \ . \tag{4.34}$$

The interchange of space and time variables in (4.30) (cf. (3.10)) means that the local wave velocity c in (3.16) has been replaced by a local "slowness" $= (\beta v)/c_0^2$ in (4.34).

Before discussing the analytical development of the solution (4.33) and (4.34) we shall treat it graphically. From (4.33) and (4.34) v is implicitly given as

$$v = v_0 \sin \omega(\tau + \frac{\beta v}{c_0^2} x) . \tag{4.35}$$

The graphical representation of (4.35) is made easier if the equation is rewritten as

$$\omega\tau = \arcsin(\frac{v}{v_0}) - \sigma\frac{v}{v_0} , \tag{4.36}$$

where σ is defined as

$$\sigma = \frac{\omega\beta v_0}{c_0^2} x . \tag{4.37}$$

The function $\omega\tau(v/v_0)$ is studied graphically in Figure 4.5 (Rudenko and Soluyan 1977)

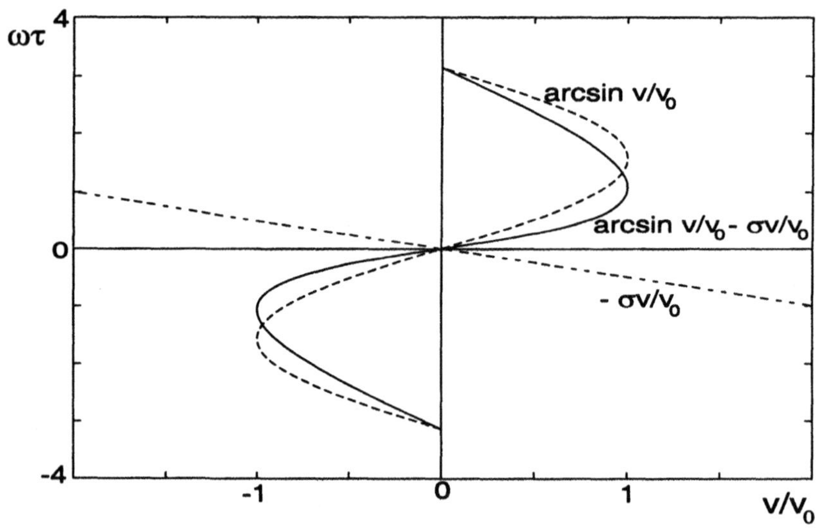

Figure 4.5: Graphical representation of (4.36).

The slope of the straight line $-\sigma v/v_0$ increases in magnitude for increasing σ values, until a discontinuity in the function $v(\omega\tau)/v_0$ begins to develop. This occurs for $\sigma = 1$, because the derivative at the origin of $\omega\tau$ with respect to v/v_0 becomes zero for $\sigma = 1$. The discontinuous solution for $\sigma > 1$ will be studied later in this section.

4.2.2 The Bessel-Fubini solution

We will now derive a more explicit expression for the solution (4.35), the Bessel-Fubini formula.

The implicit dependence of v/v_0 on $\omega\tau$ can be explicitely given as a Fourier series

$$\frac{v}{v_0} = \sum_{n=1}^{\infty} B_n(\sigma) \sin n\omega\tau \ , \tag{4.38}$$

where the Fourier coefficients $B_n(\sigma)$ are given as

$$B_n = \frac{2}{\pi} \int_0^{\pi} \frac{v}{v_0} \sin n\omega\tau \, d(\omega\tau) \ . \tag{4.39}$$

In the integral (4.39) we make the substitution

$$\xi = \omega\tau + \sigma\frac{v}{v_0} \ , \tag{4.40}$$

which, in view of (4.35) and (4.37), gives

$$\frac{v}{v_0} = \sin\xi \ . \tag{4.41}$$

From (4.40) and (4.41) we obtain

$$\omega\tau = \xi - \sigma\sin\xi \ . \tag{4.42}$$

Integrating (4.39) partially and using (4.41) and (4.42) we obtain

$$\begin{aligned} B_n &= \frac{2}{\pi} \int_0^{\pi} \sin\xi \sin n\omega\tau \, d(\omega\tau) \\ &= -\left[\frac{2}{n\pi} \sin\xi \cos n\omega\tau\right]_{\omega\tau=0}^{\omega\tau=\pi} + \frac{2}{n\pi} \int_0^{\pi} \cos\xi \cos[n(\xi - \sigma\sin\xi)] \, d\xi \\ &= \frac{2}{n\pi} \int_0^{\pi} \frac{1}{2} \{\cos[(n+1)\xi - n\sigma\sin\xi] + \cos[(n-1)\xi - n\sigma\sin\xi)]\} \, d\xi \\ &= \frac{1}{n}[J_{n+1}(n\sigma) + J_{n-1}(n\sigma)] = \frac{2}{n\sigma} J_n(n\sigma) \ , \end{aligned} \tag{4.43}$$

where we have employed the wellknown integral representation of a Bessel function of integer order

$$J_n(z) = \frac{1}{\pi} \int_0^{\pi} \cos(n\xi - z\sin\xi) d(\xi) \tag{4.44}$$

and the recursion relation for Bessel functions

$$J_{n+1}(z) = \frac{2n}{z} J_n(z) - J_{n-1}(z) \ . \tag{4.45}$$

The series (4.38) thus becomes

$$\frac{v}{v_0} = 2 \sum_{n=1}^{\infty} \frac{J_n(n\sigma)}{n\sigma} \sin n\omega\tau \tag{4.46}$$

and is called the Bessel-Fubini formula (Fubini-Ghiron 1935; Blackstock 1962). It is valid only when the implicit solution (4.35) is unique, i.e. for $\sigma < 1$. The transformation (4.42) directly indicates that the solution (4.46) cannot be valid for $\sigma > 1$, because then the interval $(0, \pi)$ for $\omega\tau$ does not correspond to the interval $(0, \pi)$ for ξ.

An alternative derivation of the Bessel-Fubini formula was given by Banta (1965). A modification of the Bessel-Fubini solution derived for a spherically diverging wave form was given by Cary (1971).

4.2.3 Sawtooth solution

For $\sigma > 1$ the solution (4.41) and (4.42) develops a discontinuity at $\omega\tau = 0$. If we put

$$\frac{v}{v_0} = V \ , \tag{4.47}$$

then (4.30) can be written in the dimensionless form with $\omega\tau = \theta$. We then find (2.76) with $\epsilon = 0$

$$\frac{\partial V}{\partial \sigma} - V \frac{\partial V}{\partial \theta} = 0 \ . \tag{4.48}$$

If the position of the discontinuity is (σ, θ_{sh}), then in analogy with (3.36) we have

$$\frac{d\theta_{sh}}{d\sigma} = -\frac{1}{2}(V_1 + V_2) \ , \tag{4.49}$$

where V_1 and V_2 are the values of V to the left and to the right of θ_{sh} respectively for a certain σ value. In analogy with (3.39) we then obtain using (4.42)

$$\theta_{sh} = \xi_1 - \sigma \sin \xi_1 = \xi_2 - \sigma \sin \xi_2 \tag{4.50}$$

or

$$\theta_{sh} = \arcsin V_1 - \sigma V_1 = \arcsin V_2 - \sigma V_2 \ . \tag{4.51}$$

For $\theta_{sh} = 0$ we obtain from (4.49)

$$V_1 = -V_2 \ , \tag{4.52}$$

where, according to (4.51), the amplitude V_2 is determined by the equation

$$V_2 = \sin(\sigma V_2) \ . \tag{4.53}$$

For $\sigma < 1$ the equation (4.53) has only the trivial solution $V_2 = 0$, which means that the solution to (4.48) is continuous according to (4.52). For $\sigma > 1$, the shock amplitude V_2 is greater than zero. To obtain an analytical solution we must assume $\sigma \gg 1$, under which assumption we will now derive a discontinuous solution. If we put

$$p = \sigma V_2 \ , \tag{4.54}$$

equation (4.53) becomes

$$p = \sigma \sin p \ . \tag{4.55}$$

We see from equation (4.55) that $\sin p \ll 1$ for $\sigma \gg 1$; consequently p is close to π. Writing

$$p = \pi - \delta \qquad \delta \ll 1 \ , \tag{4.56}$$

equation (4.55) can be written

$$\pi - \delta = \sigma \sin \delta \approx \sigma\delta \ . \tag{4.57}$$

Therefore

$$\delta = \frac{\pi}{1+\sigma} \ , \qquad \sigma \gg 1 \tag{4.58}$$

$$V_2 = \frac{\pi - \delta}{\sigma} = \frac{\pi}{1+\sigma} \ , \qquad \sigma \gg 1 \ . \tag{4.59}$$

From (4.40), (4.41) and (4.47) we obtain V as an implicit function of θ:

$$V = \sin(\theta + \sigma V) \ . \tag{4.60}$$

The function $V(\theta)$ may for large σ be approximated by a straight line. This line is different to the left ($-\pi < \theta < 0$) and to the right ($0 < \theta < \pi$) of the discontinuity. Putting

$$V = \frac{\pi - \theta}{1 + \sigma}, \qquad 0 < \theta < \pi, \ \ \sigma \gg 1 \tag{4.61}$$

in (4.60) we find that the righthand side becomes $\sin(\theta + \sigma\frac{\pi-\theta}{1+\sigma}) = \sin(\pi - (\theta + \sigma\frac{\pi-\theta}{1+\sigma})) = \sin((\pi - \theta)/(1+\sigma)$. Thus the approximation becomes better the more θ approaches π. Analogously the linear approximation for V to the left of the discontinuity

$$V = -\frac{\pi + \theta}{1 + \sigma}, \qquad -\pi < \theta < 0, \ \ \sigma \gg 1 \tag{4.62}$$

is valid. The expressions (4.61) and (4.62) describe a fully developed sawtooth wave, illustrated in Figure 4.6. For $\sigma > 1$ but not approaching infinity the sawtooth wave is not fully developed, but some structure still remains on the slopes, while the shock always is discontinuous.

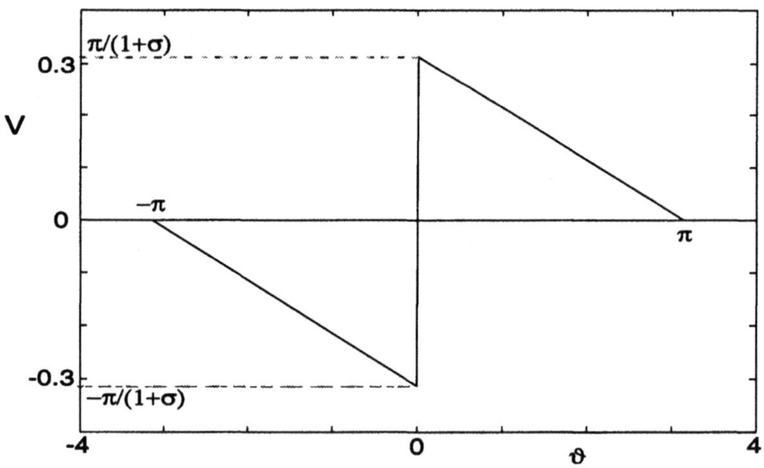

Figure 4.6: A period of the fully developed sawtooth wave.

For the not fully developed sawtooth wave we can now represent the wave as a Fourier expansion analogous to the Bessel-Fubini expansion (4.46). To that end we calculate the Fourier coefficients $B_n(\sigma)$ in the expansion (4.38). Using (4.41) we obtain first in analogy with the case $\sigma < 1$ (see (4.43))

$$B_n = \frac{2}{\pi}\int_0^\pi V \sin n\theta \, d\theta = \frac{2}{\pi}\int_0^\pi \sin\xi \sin n\theta \, d\theta$$
$$= \frac{2}{n\pi}[-\sin\xi\cos n\theta]_{\theta=0}^{\theta=\pi} + \frac{2}{n\pi}\int_{\theta=0}^{\theta=\pi} \cos n\theta \cos\xi \, d\xi \ . \quad (4.63)$$

In the case $\sigma > 1$ we find from (4.42) and (4.53) for the function $\xi(\theta)$:

$$\xi(\pi) = \pi \quad (4.64)$$
$$\xi(0) = \xi_{min} = \sigma V_2 \ . \quad (4.65)$$

Using (4.64), (4.65) and (4.42) in (4.63) we obtain

$$B_n = \frac{2}{n\pi}\sin\xi_{min} + \frac{2}{n\pi\sigma}\int_{\theta=0}^{\theta=\pi} \cos n\theta (d\xi - d\theta) \ . \quad (4.66)$$

Since only the integral over $d\xi$ contributes in (4.66), we obtain through using (4.42) the expression (Blackstock 1966)

$$B_n = \frac{2}{n\pi}V_2 + \frac{2}{n\pi\sigma}\int_{\xi_{min}}^{\pi} \cos n(\xi - \sigma\sin\xi) d\xi \ , \quad (4.67)$$

where V_2 is given by (4.53) and ξ_{min} by (4.65).

From (4.59) it is seen that because of the limit

$$\xi_{min} \longrightarrow \pi \qquad \text{as } \sigma \to \infty \ , \quad (4.68)$$

we obtain

$$B_n = \frac{2}{n\pi}V_2 = \frac{2}{n(1+\sigma)} \ , \quad (4.69)$$

which are the wellknown coefficients of the Fourier expansion of the sawtooth solution (4.61) and (4.62). The expression (4.67) gives an evaluation of the Fourier coefficients of the expansion (4.38) in the whole σ region where both the sawtooth term (the first term) and the Bessel-Fubini term (the second term) are important.

Many aspects of sawtooth waves were discussed in the review article by Rudenko (1995). Sawtooth waves in different geometries were studied by simpler mathematical methods by Temkin (1969[a,b]). Random sawtooth waves were discussed by Rudenko and Khokhlova (1991).

4.2.4 The one saddle-point method

Now a method for solving the equation (4.48) will be presented whose result is differing in form while numerically equivalent to the one in (4.67). It is based in the saddle point method (see page 50).

Below are presented two equivalent variations of the single frequency boundary condition, which differ only by a shift in time θ where the function F is obtained through (3.90):

$$V_0 = \sin\theta \quad \Rightarrow \tag{4.70}$$

$$F(\theta') = \frac{1}{2\epsilon}(-\cos\theta' - \frac{(\theta-\theta')^2}{2\sigma}) \tag{4.71}$$

$$V_0 = -\sin\theta \quad \Rightarrow \tag{4.72}$$

$$F(\theta') = \frac{1}{2\epsilon}(\cos\theta' - \frac{(\theta-\theta')^2}{2\sigma}) . \tag{4.73}$$

Since the dissipation over non-linearity ratio ϵ is a small number ($\epsilon \ll 1$) the exponent F will have marked maxima. This property is well suited for the saddle-point method, see page 50, based on the fact that the main contributions to such integrals are from areas around the maxima of the exponent. From (4.71) we observe that we have two maxima symmetrically around $\theta' = 0$, but for (4.73) there is one maximum *at* $\theta' = 0$, see Figure 4.7.

When $V_0 = \sin\theta$ two different maxima are needed as evident from the appearance of a shock at $\theta = 0$ after a certain distance. For $V_0 = -\sin\theta$ the shocks are at $\pm\pi$ and only one maximum is needed to describe the interval within $[-\pi : \pi]$. But the endpoints are not in this interval and the result is simply a continuous line without shocks. There is only one maximum per period.

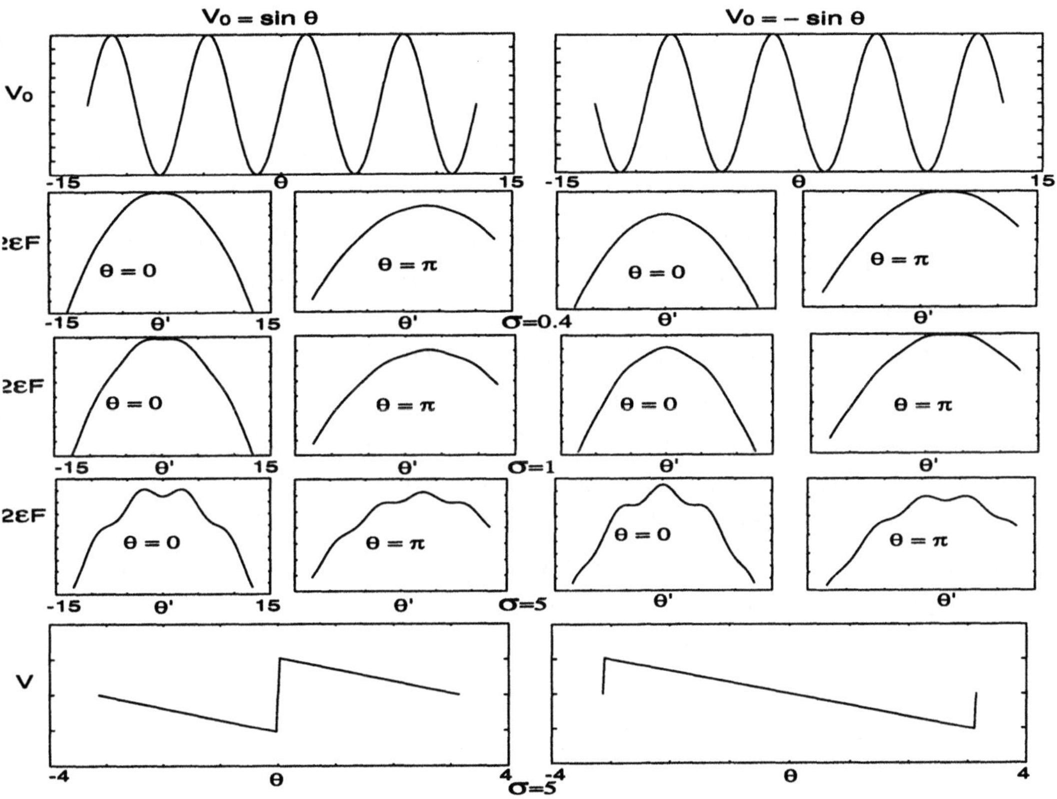

Figure 4.7: A comparison of the exponents F for $V_0 = \sin\theta$ and $V_0 = -\sin\theta$ (normalized for any constant ϵ). From equations (4.71) and (4.73).

The saddle-point is the maximum point of F. The condition for maxima is

$$F'(\theta_0) = 0 \ , \tag{4.74}$$

which gives from (4.71) and (4.73)

$$\sin\theta_0 + \frac{\theta - \theta_0}{\sigma} = 0 \ , \tag{4.75}$$

$$-\sin\theta_0 + \frac{\theta - \theta_0}{\sigma} = 0 \ . \tag{4.76}$$

The series expansion around θ_0 is

$$F(\theta') \sim F(\theta_0) + \frac{(\theta_0 - \theta')^2}{2} F''(\theta_0) \ . \tag{4.77}$$

There are for large σ plenty of maxima of $F(\theta')$ in (4.71) and (4.73) in the interval $[-\infty, \infty]$ but as the square increases the further we get from θ, which is in the interval $[-\pi, \pi]$, only maxima close to 0 are of interest. This is counteracted by the fact that σ is growing, making the square decrease.

If the dissipation ϵ is not equal to zero there exists in principle information about the entire signal even after the formation of shock too. Unfortunately, it is difficult to extract this information. To retain enough information about the wave after shock formation in principle all saddle-points must be included.

No part of a wave can be described by only one saddle-point when the dissipation is not equal to zero. All saddle-points influence all parts of the wave to different degrees. Therefore the term in the solutions containing the dissipation parameter ϵ, is not expected to be correct. Indeed it is not, and will be written only in the first instance and ignored later.

In conclusion, only the single frequency solution may be adequately described by one saddle-point at all distances when the dissipation is zero. Multiple frequencies may be described for zero dissipation only before shock formation.

The boundary condition [1]

$$V_0 = -\sin\theta \tag{4.78}$$

gives, through (3.89), and (4.73) the solution - ignoring the diffusion ϵ -

$$V(\sigma, \theta) = -\sin\theta_0 \tag{4.79}$$

with $\theta_0(\sigma, \theta)$ obtained from (4.76)

$$\theta_0 + \sigma\sin\theta_0 = \theta. \tag{4.80}$$

[1] We could have used $V_0 = \sin\theta$ and instead have looked at the interval $[0, 2\pi]$. The important point is to keep the shocks at the endpoints of the interval, and none inside the interval.

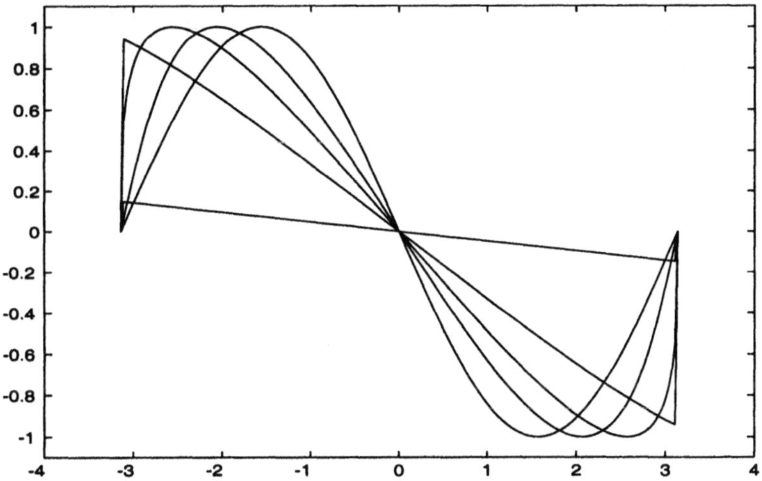

Figure 4.8: The wave developed from $V_0 = -\sin\theta$ for the following distances $\sigma = 0, 0.5, 1, 2$ and 20. From (4.79).

The equation (4.79) is the same as (4.60) and continues the zero diffusion solution into the shock region which goes on in perpetuity.

The Fourier expansion of (4.79) is

$$V(\sigma,\theta) = \sum_{m=1}^{\infty} d_m \sin m\theta \tag{4.81}$$

$$\begin{aligned} d_m &= \frac{1}{\pi}\int_0^{\theta=\pi} (-\sin\theta_0)\sin m\theta\, d\theta \\ &= -\frac{1}{\pi}\int_0^{\theta_0=\pi} \sin\theta_0 \sin m(\theta_0 + \sigma\sin\theta_0)[d\theta_0 + \sigma\cos\theta_0 d\theta_0] \quad (4.82) \\ &= -\frac{1}{\pi}[I_{m-1,m} - I_{m+1,m} + \frac{\sigma}{2}I_{m-2,m} - \frac{\sigma}{2}I_{m+2,m}]\,, \quad (4.83) \end{aligned}$$

where $I_{n,m}$ are incomplete modified Bessel functions

$$\begin{aligned} I_{n,m} &= \int_0^{\theta=\pi} \sin\theta_0 \sin m\theta d\theta \\ &= \{d\theta = [(4.80)] = d\theta_0 + \sigma\cos\theta_0 d\theta_0\} \\ &= \int_0^{\theta_a} \cos[n\theta_0 + \sigma m\sin\theta_0]d\,\theta_0 \quad (4.84) \end{aligned}$$

The integration limit θ_a is found by solving (4.80) for the first root larger than zero.

$$\theta_a + \sigma \sin \theta_a = \theta = \pi \tag{4.85}$$

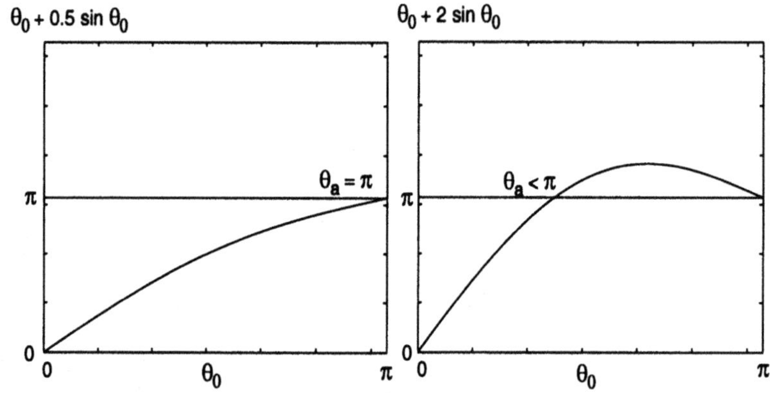

Figure 4.9: The solution of (4.85) is at the value $\theta_0 = \theta_a$ given by where the two curves intersect. While $\sigma \leq 1$, examplified by $\sigma = 0.5$ (left), we have $\theta_a = \pi$. For $\sigma > 1$, examplified by $\sigma = 2$ (right), θ_a decreases rapidly.

With growing distance, θ_a gets smaller, see Figure 4.9.

There is no theoretical restriction on the valid σ-region of this solution, except in reality there is of course always dissipation. In Figure 4.10 is seen the evolution of the first three harmonics.

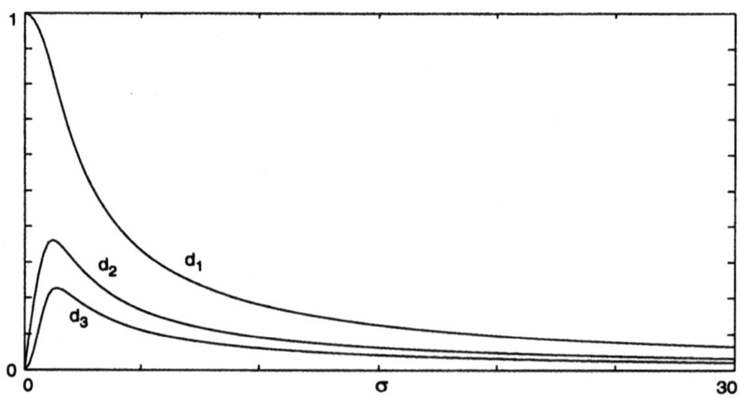

Figure 4.10: The absolute value of the first three harmonics for the original signal $V_0 = -\sin \theta$.

For $\sigma < 1$ the coefficient is simplified to

$$d_m = (-1)^m \frac{2}{m\sigma} J_m(m\sigma) \tag{4.86}$$

This is the Bessel-Fubini coefficient, B_m in (4.46), which was obtained from the Riemann equation. The $(-1)^m$ is the difference between the plus and minus sine. This may be explained by that the change from $\sin\theta$ into $-\sin\theta$ is the same as shifting the time axis by π. Then $\sin(m(\theta+\pi)) = \sin(m\theta)\cos(m\pi) + \sin(m\pi)\cos(m\theta) = \sin(m\theta)(-1)^m$.

stötpunkterna beskriven av dess egen enda

How is it possible to obtain the Fourier series explicitly only for one frequency when using one saddle-point ? In Figure 4.11 is seen how the time domain solution for one saddle-point after shock formation continues past the interval endpoints $-\pi$ and π.

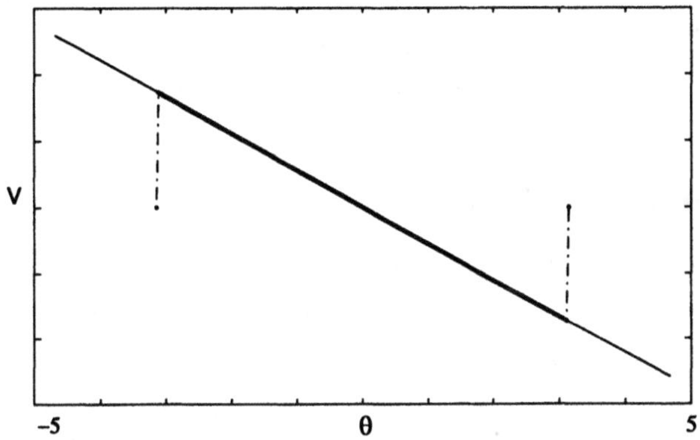

Figure 4.11: The result when using the same saddle-point: the wave continues as a straight line past the true shocks. The distance $\sigma = 8$.

The Fourier expansion of this straight line thus contains frequencies which are lower than the lowest frequency of the problem. This is consequently not a solution to the problem. But for a single frequency the period is known and the solution adapts to the true lowest period by imposing a Fourier serie ansatz. For more complicated waves this is not possible, all that is accomplished is an adaption to the period of the total wave.

For large distances the amplitude of the total wave is, from (4.79) with $\sin\theta_0 \to \theta_0$ and $\cos\theta_0 \to 1$,

$$V(\sigma,\theta) \approx -\frac{\theta}{1+\sigma} \tag{4.87}$$

This is the saw-tooth shape found also in (4.61), (4.62).

For large distances when $\sigma \gg 1$, d_m may be approximated into a sawtooth shape with coefficients

$$d_{m\mathrm{sawtooth}} = (\pm 1)^m \frac{2}{m(1+\sigma)}, \tag{4.88}$$

The plus is for $V_0 = \sin\theta$, and the minus for $V_0 = -\sin\theta$, corresponding to the limit of (4.81).

Examples of other use of the saddle-point method with $\epsilon \to 0$ are found in the book by Gurbatov, Malakhov and Saichev (1991) and in the paper by Gurbatov, Enflo and Pasmanik (1999).

4.2.5 Time reversal

In linear time reversed acoustics a signal is received, and the timereversed signal is retransmitted back in the direction from which it came. This simple technique turns out to have several applications in medicine, underwater communication, hydrodynamics and material analysis (Fink 1997). Linear time reversal experiments show how multiple scattering media are better at refocusing than homogeneous as the effective focusing aperture is wider, and that the method is also extraordinary insensitive to initial conditions (Derode, Roux and Fink 1995). As a way of examining objects, e.g. in underwater (Kuperman, Hodgkiss, Song, Akal, Ferla and Jackson 1998) or medical ultrasound, a three-step sequence gives the illumination of specific point reflectors. First a pulse is sent and reflected from a point. This signal is then received by an array of transducers, time reversed and reemitted. This new pulse will focus on the point reflector, automatically compensating for irregularities (refraction, mode conversion and anisotropy) in the medium, and thus giving a strong response (Dorme and Fink 1995).

The time invariant property of linear, non-dissipative waves has been emphasized. This invariant property is manifested when a wave is received, time reversed, retransmitted and the nearly original wave is received (Jackson and Dowling 1991) at the point of the transmission of the original signal. This time reversal invariance is violated when dissipation is taken into account and certainly to a higher degree when nonlinear effects are present. The signal received at the point of transmission will not be the original signal.

Nonlinear time reversal was presented by Hedberg (1997), experimentally demonstrated by Brysev, Bunkin, Hamilton, Krutyanski, Cunningham, Preobrazhenskii, Pylnov, Stakhovskii and Younghouse (1998) and numerically simulated by Cunningham, Hamilton, Brysev and Krutyansky (2001).

In a second form time reversal can be used when trying to answer the question of what one can know about the original wave when measuring a specific signal.

These are not only one kind of time reversal. In the first case, time reversal means that a measured signal is re-transmitted through the transducers in opposite order in time - first in is last out. This is time reversal in the form of reversal of a vector (the measured signal) in a computer. As this is used in reality time itself is of course never reversed. The signal propagating out from the transducers focus on approximately the same spot as it originated through almost any kind of diffractive interferences. Nonlinearity and dissipation will distort the signals in a way from which they can not recover in the propagation from the transducers.

In the second case trying to find the original wave, calculations are done backwards in time. We are letting time go in negative direction in the solving of equations. This can in theory compensate for the both the nonlinear and dissipative degradation, of course as well as the diffraction. We will back-propagate using time going in the negative direction and will inside a limited propagation region get back the original signal. Outside this limited region information will be lost from the signal and the original signal may be recovered only through making some assumptions.

In the description which follows we will look at one-dimensional nonlinear propagation. Neither diffraction nor dissipation will be taken into consideration. We will investigate the influence of purely nonlinear distortion on signals and what is possible to ascertain about the origin of the signals.

Here will be shown at which distances a measured wave can have been transmitted when having propagated through a dissipation free and nonlinear homogeneuos medium. If one makes the assumption that the wave was harmonic to begin with, the point of origin can be calculated. In Figures 4.12 and 4.13 are explained by graphical representations what can be deduced from nonlinear time reversal. The distance $\sigma = 0$ is defined as where the wave would have been sinusoidal. When measuring the wave at a certain point, while not knowing the origin, we can say that the wave must have been transmitted at a distance which is at its longest the point where it would be going into a shock if travelling backwards. One can not say more as the wave could have been transmitted as it is just before measuring.

In Figure 4.12 the measured wave is not in shock and the backpropagation will give the exact wave form of the original wave.

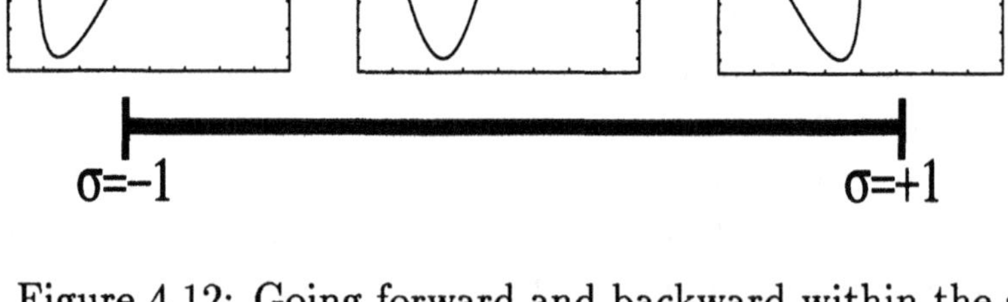

Figure 4.12: Going forward and backward within the shock formation distance the wave is always the same for a specific point. The interval over which the Bessel-Fubini solution may be used is the same interval that is continuous for a nondissipative wave.

When entering the discontinuous region, Figure 4.13, the original wave will not be recreated through backpropagation. Within the interval that the wave had propagated at its farthest, the wave may be going back towards $\sigma = -1$ (just as in Figure 4.12) and forward again without losing *more* of its original shape. The wave has lost part in a shock and the backpropagation gives a wave which has a straight line, within which region nothing can be extracted to tell us about the original wave. It could have been as it looks in the figure keeping the straight line, it could have had a spike of any height, or it could perhaps have been sinusoidal. If making an assumption of the original shape of the wave - for example, assume that it was sinusoidal - then the whole part of the wave can be recreated exactly from the parts which has not been in shock. It is really so that, if the wave can be described by a continuous function, these non-shocked parts contain in theory information about the total original wave. These parts shrink with increase in distance after shock. And of course, in reality attenuation, and limited measurement accuracy, will make these parts give almost no practical information for large distances.

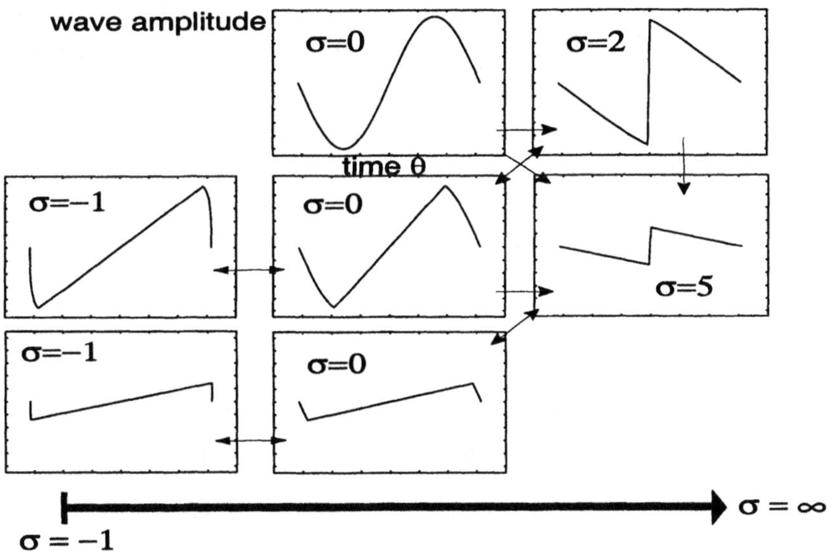

Figure 4.13: Going forward: The positive slopes of the wave steepens and goes towards shock.
Going backwards: the negative slopes steepen and go towards shocks which limits the possible maximum distance which the wave may have travelled, which for this single sine is the distance $\sigma = -1$.

4.3 Modulated Riemann waves

To go from a boundary condition with a single frequency to multiple frequencies complicate matters. There does not exist the same short periodicity, if any, to help finding the solution. We get shocks with different and most often asymmetric amplitudes. When a shock amplitude is asymmetric the shock will move on the retarded time axis. Several shocks will be moving with different velocities. The frequency spectrum is at distances near the source very complicated because of continuous creation of sums and differences of existing frequencies. This means that both higher and lower harmonics are created. The number of shocks decrease as they collide and merge into each other.

If the wave is periodic there will at large distances finally exist a wave with often only one shock per period. Then it will in practice (but not in theory) be impossible to see if the wave has emerged from a simple or a complicated boundary condition. If one shock is left, the shape will be a regular sawtooth wave. It could have come from a single sine wave, from a bifrequency sine wave with both frequencies higher harmonics of the final period or any other wave with the same periodicity. For some periodic boundary conditions two or more shocks will be sustained within the period.

If the wave is not periodic, e.g. two frequencies whose ratio is incommensurable, there will forever be shocks moving relatively each other.

In this section will be discussed mainly the simplest of the multi-frequency boundary condition - the two-frequency - where most phenomena pertaining to multiple frequencies are exhibited. We study the evolution of a wave with the boundary condition

$$V(\sigma = 0, \theta) = V_0 = a_1 \sin(b_1\theta + \phi_1) + a_2 \sin(b_2\theta + \phi_2)\ . \qquad (4.89)$$

4.3.1 Direct method for bifrequency boundary condition

First in this section a generalization of the Bessel-Fubini solution (Fubini-Ghiron 1935) valid before the formation of shock will be derived. Substitutions are made with $b_1\theta = u$ and $b_2\theta = v$ in (4.89) (let

$\phi_1 = 0$ and $\phi_2 = 0$ for now) so that

$$V_0(u,v) = a_1 \sin u + a_2 \sin v = \mathrm{Im}\{a_1 e^{iu} + a_2 e^{iv}\}\ . \tag{4.90}$$

A solution in the form of a double Fourier series is

$$V(\sigma,\theta) = \sum_{n=-\infty}^{\infty} \sum_{m=-\infty}^{\infty} c_{mn} \exp[i(nu+mv)]\ , \tag{4.91}$$

with

$$c_{mn} = \frac{1}{4\pi^2} \int_{-\pi}^{\pi} \int_{-\pi}^{\pi} V(\sigma,u,v) \exp[-i(nu+mv)]\, du\, dv\ . \tag{4.92}$$

Change variables from $(u = b_1\theta, v = b_2\theta)$ to $(\mu = b_1\theta_0, \nu = b_2\theta_0)$ according to

$$\begin{aligned} u &= b_1\theta = \{(3.57)\} = b_1(\theta_0 - \sigma V_0(\mu,\nu)) \\ &= \mu - b_1\sigma \mathrm{Im}\{a_1 e^{i\mu} + a_2 e^{i\nu}\} \end{aligned} \tag{4.93}$$

$$\begin{aligned} v &= b_2\theta = \{(3.57)\} = b_2(\theta_0 - \sigma V_0(\mu,\nu)) \\ &= nu - b_2\sigma \mathrm{Im}\{a_1 e^{i\mu} + a_2 e^{i\nu}\}\ . \end{aligned} \tag{4.94}$$

The differential product gives

$$\begin{aligned} du\, dv &= \begin{vmatrix} \dfrac{\partial u}{\partial \mu} & \dfrac{\partial u}{\partial \nu} \\ \dfrac{\partial v}{\partial \mu} & \dfrac{\partial v}{\partial \nu} \end{vmatrix} d\mu\, d\nu \\ &= 1 - \sigma[b_1 a_1 \mathrm{Im}(ie^{i\mu}) + b_2 a_2 \mathrm{Im}(ie^{i\nu})]\, d\mu\, d\nu\ . \end{aligned} \tag{4.95}$$

Changing the integration limits is simple as long as $\theta_0(\theta)$, and thus $\mu(u)$ and $\nu(v)$, are single valued functions:

$$(u = \pi, v = \pi) \rightarrow (\mu = \pi, \nu = \pi) \tag{4.96}$$

$$(u = -\pi, v = -\pi) \rightarrow (\mu = -\pi, \nu = -\pi)\ . \tag{4.97}$$

These transforms are valid only as long as the single value condition is satisfied.

Still when single valued, it is fortunately possible to proceed as the integrals are of the Bessel function form (Hochstadt 1971, p. 207):

$$J_n(z) = \frac{1}{2\pi}\int_{-\pi}^{\pi} e^{i(z\sin\gamma - n\gamma)}\, d\gamma \ . \qquad (4.98)$$

These Bessel functions may be combined through the following two relations: (Hochstadt 1971, pp. 208-209)

$$J_{-n}(z) = (-1)^n J_n(z) \qquad (4.99)$$

$$\frac{2n}{z} J_n(z) = J_{n+1}(z) + J_{n-1}(z) \ . \qquad (4.100)$$

The final result, first calculated by Fenlon (1972) for a general multi-frequency source, is

$$V(\sigma,\theta) = \sum_{m=-\infty}^{\infty}\sum_{n=-\infty}^{\infty} \frac{J_m(a_1\sigma(mb_1+nb_2))J_n(a_2\sigma(mb_1+nb_2))}{i(mb_1+nb_2)\sigma} \times \exp[im(b_1\theta+\phi_1)+in(b_2\theta+\phi_2)] \ . \qquad (4.101)$$

This solution is a generalization of the Bessel-Fubini solution (4.46) for a single sine wave. It is valid only up to the point (Cary 1973)

$$\sigma_{sh} = \frac{1}{a_1 b_1 + a_1 b_2} \quad (\text{when } \phi_1 = \phi_2 = 0) \ , \qquad (4.102)$$

where (4.93) and (4.94) become multi-valued and which physically is the distance where the first shock appears. This changes the integration intervals in the Fourier integral and Bessel functions are no longer representative.

As an example the main contribution [2] to the difference frequency $\Delta = b_1 - b_2$ from (4.101) is

$$V_\Delta = \frac{2}{\Delta\sigma} J_1(a_1\Delta\sigma) J_1(a_2\Delta\sigma)\sin(\Delta\theta) \qquad (4.103)$$

When the difference frequency is much smaller than the original frequencies, $\Delta << b_1, b_2$, the growth is in the beginning almost linear:

$$V_\Delta = \frac{a_1 a_2 \Delta\sigma}{2}\sin(\Delta\theta) \ . \qquad (4.104)$$

[2] There exist an infinite number of higher order contributions where $nb_1 + mb_2 = \Delta$ (n and m are integers).

4.3.2 The one saddle-point method for bifrequency boundary condition

The one saddle-point method might be extended from one frequency (see page 66) to multiple frequencies. In this section only two frequency boundary conditions are presented. Below are presented two variations of two-frequency boundary conditions

$$V_0 = a_1 \sin b_1\theta + a_2 \sin b_2\theta \quad \Rightarrow$$
$$F(\theta') = \frac{1}{2\epsilon}(-\frac{a_1}{b_1}\cos(b_1\theta') - \frac{a_2}{b_2}\cos(b_2\theta') - \frac{(\theta-\theta')^2}{2\sigma}) \quad (4.105)$$

$$V_0 = -a_1 \sin b_1\theta - a_2 \sin b_2\theta \quad \Rightarrow$$
$$F(\theta') = \frac{1}{2\epsilon}(\frac{a_1}{b_1}\cos(b_1\theta') + \frac{a_2}{b_2}\cos(b_2\theta') - \frac{(\theta-\theta')^2}{2\sigma}) \quad (4.106)$$

For the two frequency conditions (4.105) and (4.106) there are several maxima with different values and the difference is more accentuated than for the single frequency conditions. As the distance increases, the influence of these maxima will change. For dissipation free propagation fewer and fewer maxima will carry any importance and finally one or two maxima is all that is needed to describe the wave. For the dissipation-free case it is enough to consider one saddle point at a time, because as $\epsilon \to 0$, only the largest maximum for any given θ and σ contributes to the integral. With increase in distance fewer maxima exist, and the disappearance of a maximum means that two shocks have merged. These local one saddle point descriptions are truly discontinuous, as are the shocks. In contrast, when the dissipation is not equal to zero, no part of the wave can be exactly described by one saddle-point. For the condition (4.105) the choice of maxima will depend on the parameters a_1, a_2, b_1 and b_2. For the condition (4.106) on the other hand, only one maximum is needed to describe the wave for very large distances and that is the one in $\theta = 0$. In Figure 4.14 it is seen how the wave to the left has two equally large maxima per period resulting in two shocks per period as its final shape while the negative option, to the right in the figure, gives one shock per period.

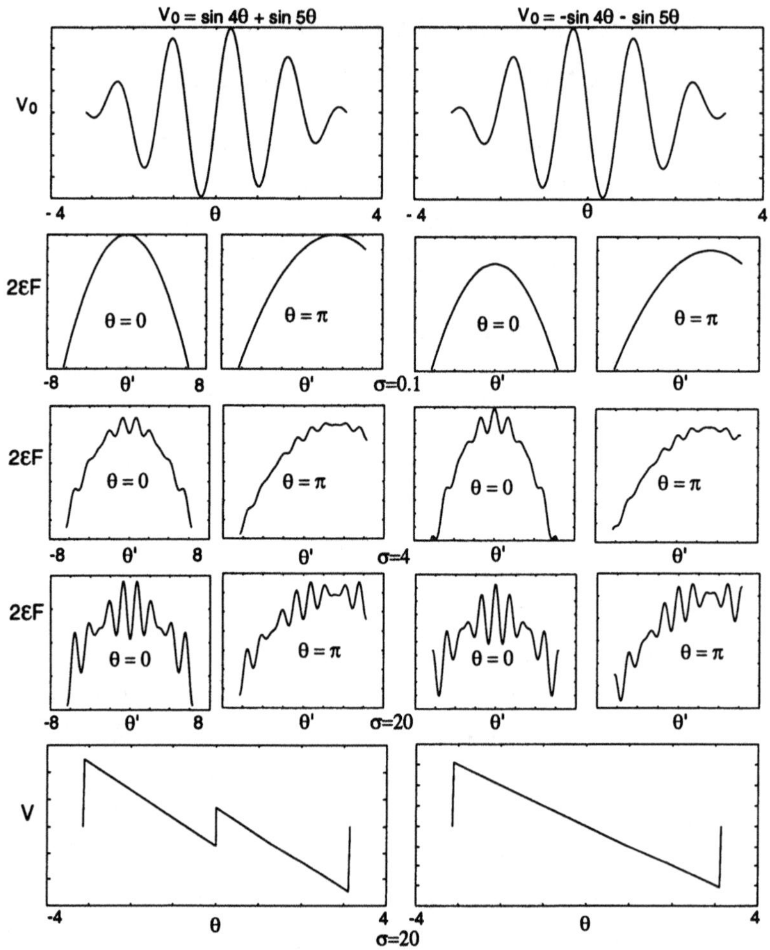

Figure 4.14: A comparison of the exponents F for $V_0 = \sin 4\theta + \sin 5\theta$ and $V_0 = -\sin 4\theta - \sin 5\theta$.

From (4.105) and (4.106) is obtained

$$a_1 \sin b_1\theta_0 + a_2 \sin b_2\theta_0 + \frac{\theta - \theta_0}{\sigma} = 0 \tag{4.107}$$

$$-a_1 \sin b_1\theta_0 - a_2 \sin b_2\theta_0 + \frac{\theta - \theta_0}{\sigma} = 0\ . \tag{4.108}$$

The boundary condition

$$V_0 = -a_1 \sin b_1\theta - a_2 \sin b_2\theta \tag{4.109}$$

gives, ignoring the dissipation and through (3.89), and (4.106), the solution

$$V(\sigma,\theta) = -a_1 \sin b_1\theta_0 - a_2 \sin b_2\theta_0 \qquad (4.110)$$

with $\theta_0(\sigma,\theta)$ obtained from (4.108)

$$\theta_0 + \sigma a_1 \sin b_1\theta_0 + \sigma a_2 \sin b_2\theta_0 = \theta \qquad (4.111)$$

The total wave, from (4.110), is shown in Figure 4.15.

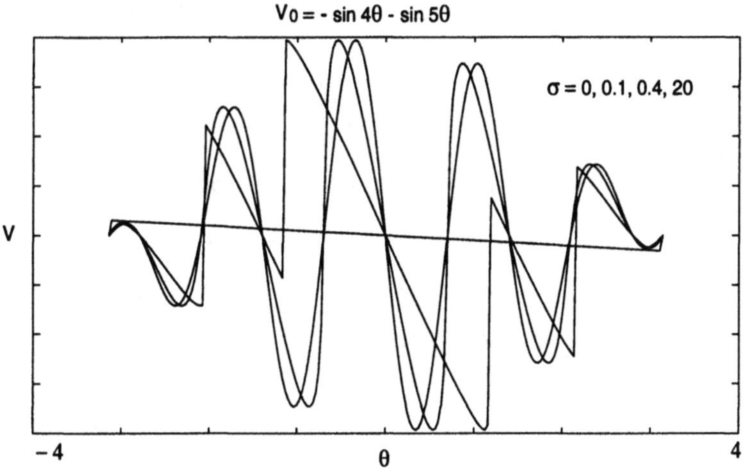

Figure 4.15: The wave developed from $V_0 = -\sin 4\theta - \sin 5\theta$, at the following distances: $\sigma = 0, 0.1, 0.4, 20$.

The Fourier expansion of (4.110) is

$$V(\sigma,\theta) = \sum_{m=-\infty}^{\infty} \sum_{n=-\infty}^{\infty} c_{mn} e^{i(mb_1+nb_2)\theta} , \qquad (4.112)$$

where the coefficients c_{mn} are given as

$$\begin{aligned} c_{mn} = &\\ & - a_1[I_{u,m-1}(Arg_u)I_{v,n}(Arg_v) + \frac{\sigma a_1 b_1}{2} I_{u,m-2}(Arg_u)I_{v,n}(Arg_v) \\ & + \frac{\sigma a_1 b_1}{2} I_{u,m}(Arg_u)I_{v,n}(Arg_v) + \frac{\sigma a_2 b_2}{2} I_{u,m-1}(Arg_u)I_{v,n-1}(Arg_v) \\ & + \frac{\sigma a_2 b_2}{2} I_{u,m+1}(Arg_u)I_{v,n-1}(Arg_v)] \end{aligned}$$

$$
\begin{aligned}
&- a_2[I_{u,m}(Arg_u)I_{v,n-1}(Arg_v) + \frac{\sigma a_1 b_1}{2} I_{u,m-1}(Arg_u)I_{v,n-1}(Arg_v) \\
&+ \frac{\sigma a_1 b_1}{2} I_{u,m+1}(Arg_u)I_{v,n-1}(Arg_v) + \frac{\sigma a_2 b_2}{2} I_{u,m}(Arg_u)I_{v,n-2}(Arg_v) \\
&+ \frac{\sigma a_2 b_2}{2} I_{u,m}(Arg_u)I_{v,n}(Arg_v)]
\end{aligned}
\tag{4.113}
$$

with the notations $I_{x,k}$, Arg_u and Arg_v explained as

$$I_{x,k}(z) = \frac{1}{2\pi} \int_{-x_a}^{x_a} \mathrm{e}^{i(z \sin\gamma - k\gamma)} d\gamma \tag{4.114}$$

$$Arg_u = -\sigma a_1(mb_1 + nb_2) \tag{4.115}$$

$$Arg_v = -\sigma a_2(mb_1 + nb_2) \ , \tag{4.116}$$

and the integration limits found by solving the following two equations for the first root larger than zero:

$$b_1\theta = \pi = u_a + \sigma a_1 b_1 \sin u_a + \sigma a_2 b_1 \sin v_a \tag{4.117}$$

$$b_2\theta = \pi = v_a + \sigma a_1 b_2 \sin u_a + \sigma a_2 b_2 \sin v_a \ . \tag{4.118}$$

Before shock formation, with the integration limits $u_a = v_a = \pi$, the coefficients c_{mn} can be written as

$$c_{mn} = \frac{J_m(a_1\sigma(mb_1 + nb_2))J_n(a_2\sigma(mb_1 + nb_2))}{i\sigma(mb_1 + nb_2)} \ . \tag{4.119}$$

This is again the solution (4.101) Fenlon (1972) derived for multiple original frequencies with phases. See Fenlon (1973[a]) for the general expression describing multiple frequencies with phases.

As this result is also the exact solution to the problem with zero dissipation, a straightforward way to include phases is to insert them in the boundary condition $V_0 = -a_1 \sin(b_1\theta + \phi_1) - a_2 \sin(b_2\theta + \phi_2)$ resulting in the replacement of $b_1\theta$ and $b_2\theta$ with $b_1\theta + \phi_1$ and $b_2\theta + \phi_2$ respectively in the solution (4.119).

Unfortunately the coefficients c_{mn} (4.113) are not valid once a state of shock has been entered. The wave has become discontinuous, and

the single saddle-point used in the calculation now describes only the local wave, in time θ, in between the closest shocks. The coefficients are not the same within different continuous segments and the total wave can not be represented without having the positions of the changes of saddle-points. For every part of the wave where a shock is formed one saddle-point more is needed. This can not be adequatly described by omitting a single one - then one local period of the wave is lost and with it one shock.
At large distances the solution V_{2F_q} to (4.48) with the boundary condition (4.109) is more simple:

$$\lim_{\sigma\to\infty} V_{2F_q} = \lim_{\sigma\to\infty}\{-\frac{a_1b_1\theta}{1+\sigma(a_1b_1+a_2b_2)} - \frac{a_2b_2\theta}{1+\sigma(a_1b_1+a_2b_2)}\} \approx -\frac{\theta}{\sigma}. \tag{4.120}$$

This may be compared with the asymptotic solution (4.87) for one frequency which is $-\frac{\theta}{1+\sigma}$. These asymptotics are valid locally for some time until all the shocks have merged and one or two of the saddle-points have taken over the whole period. The positions of there saddle-points are not automatically known for the initial condition $V_0 = a_1 \sin b_1\theta + a_2 \sin b_2\theta$, but for the condition $V_0 = -a_1 \sin b_1\theta - a_2 \sin b_2\theta$ we know that the saddle-point which will be dominant is the one at $\theta = 0$. From this point in time θ the shocks will move and eat up other shocks. Of course these two boundary conditions are not equivalent, they describe different signals. (It is not like putting the minus sign in front of a single sine, $+\sin$ and $-\sin$ are just shifted in time.)

The final sawtooth is obtained when the distance σ is very large and all the shocks have merged. Before this there is another multi-sawtooth stage where the parts of the wave are locally described by sawtooth shapes.

After shock formation, an arbitrary wave can not be described by a single saddle-point. The properties of the single sine wave support the argument that, after shock formation, the wave cannot be described by higher harmonics of the lowest frequency in the problem. One must use even lower frequencies, and the wave has lost its periodicity, which is obviously incorrect. Therefore the wave must use more than one saddle-point to correctly describe shocks. The exception is precisely a single sine wave in the form of a Fourier series, where the whole period of the wave, except the shocks, can be described by one saddle-point. This is because the only period of the wave is the total period, refering to the lowest frequency in the problem.

A solution to the lossless bifrequency problem without use of Bessel functions was given by Naugolnykh, Soluyan and Khokhlov (1963[b]).

4.3.3 Characteristic multifrequency waves

Multifrequency conditions show interesting behaviour due to the complex interactions taking place between the frequencies. The Fenlon solution (4.101) is valid only up to the formation of shocks which limits its usefulness. Still there is a situation where an interesting phenomenon, called suppression of sound by sound presents itself. The boundary condition is one low frequency, $b_1 = 1$ with $a_1 = 1$, and a weak high frequency, $b_2 \gg b_1 = 1$ with $a_2 \ll a_1 = 1$:

$$V_0 = \sin\theta + a_2 \sin b_2\theta \ . \tag{4.121}$$

The largest contribution to frequency b_2 is given by adding the two coefficients $c_{m=0,n=1}$ and $c_{m=0,n=-1}$ plus $c_{m=b_2,n=0}$ and $c_{m=b_2,n=0}$ (if b_2 is an integer, otherwise there can be no contribution) from (4.101) resulting in

$$V_{Frequency=b_2} = \frac{2}{b_2\sigma}[J_0(b_2\sigma)J_1(a_2b_2\sigma) + J_{b_2}(b_2\sigma)J_0(a_2b_2\sigma)]\sin b_2\theta \ , \tag{4.122}$$

as seen in Figures 4.16 and 4.17 where the amplitude of the high frequency b_2 is plotted. It is oscillating due to its interaction with the low frequency. At some distances the high frequency's amplitude equals zero. The high frequency is at these points completely suppressed. The sudden increases in the amplitude of frequency b_2 is due to the apperance of higher harmonics of the low frequency $b_1 = 1$ whose amplitudes grow and start to dominate over the original amplitude of b_2. The amplitudes of the higher harmonics of the low frequency (which has a large amplitude in the boundary condition (4.121)) are fairly large. At some point the high harmonic of the low frequency at the frequency of the initial high frequency will become larger than the part of the high frequency coming from the initial one.

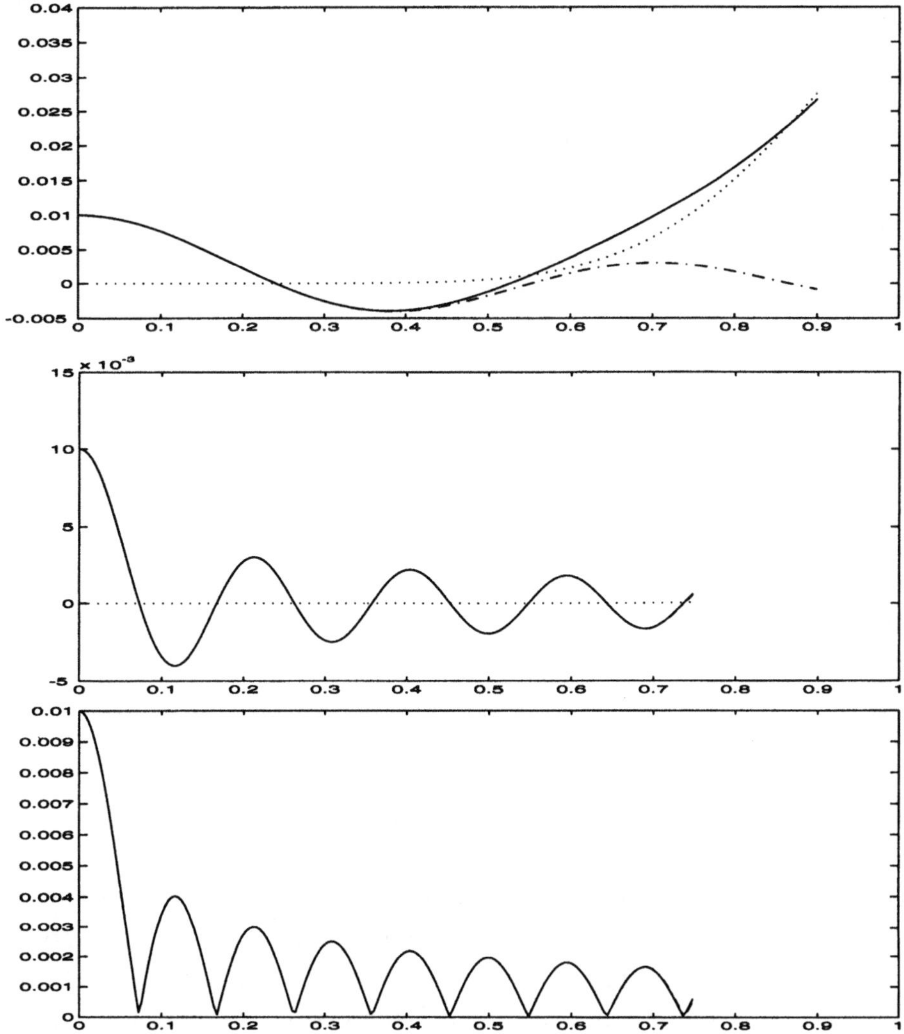

Figure 4.16: The amplitude of frequency b_2 for $V_0 = \sin\theta + a_2 \sin b_2\theta$ Top: $a_2 = 0.01$ and $b_2 = 10$; middle: $a_2 = 0.01$ and $b_2 = 33$; and bottom: The absolute value of the middle plot, which is usually how the oscillatory behaviour is shown.
Dash-dotted: the first term in (4.122). Dotted: the second term in (4.122). Solid: both terms.
Increasing the higher frequency, the interval for which the Fenlon solutions (4.101) and (4.122) are valid decreases slightly, but several more oscillations fit into the interval.

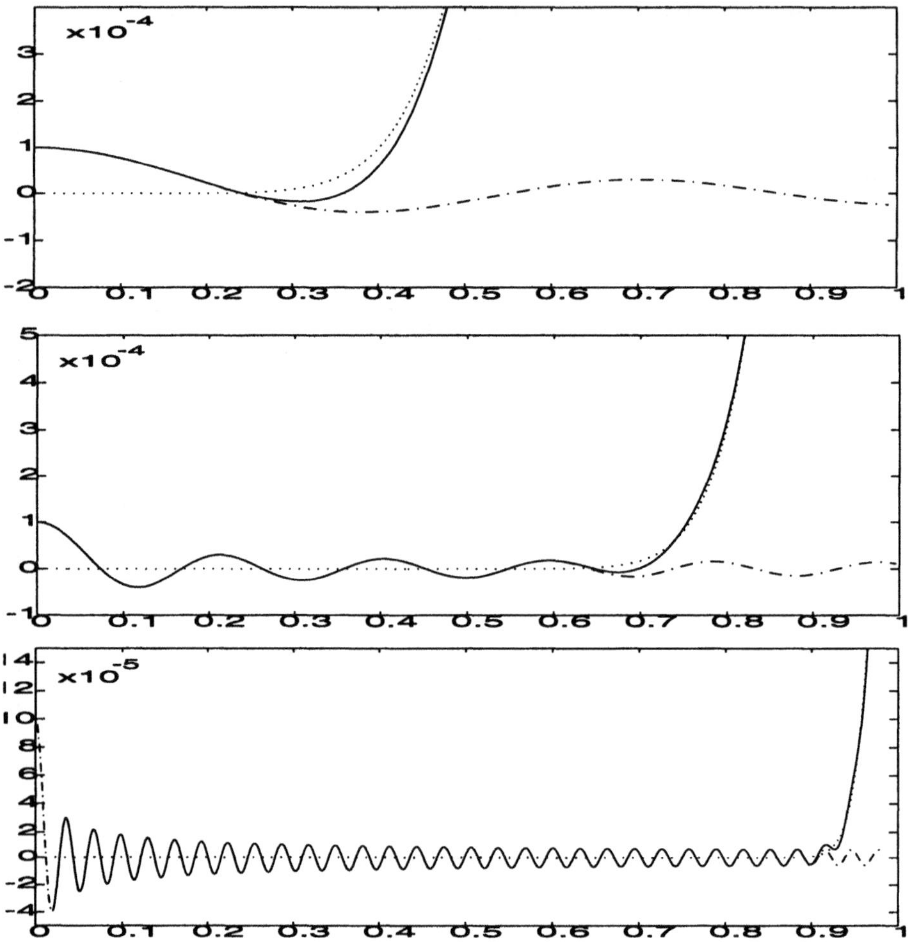

Figure 4.17: The amplitude of frequency b_2 for $V_0 = \sin\theta + a_2 \sin b_2\theta$ Top: $a_2 = 0.0001$ and $b_2 = 10$; Middle: $a_2 = 0.0001$ and $b_2 = 33$; and Bottom: $a_2 = 0.0001$ and $b_2 = 200$ Dash-dotted - the first term in (4.122); dotted - the second term in (4.122); solid - both terms. A decreased amplitude of the higher harmonic makes the higher harmonic of the lower frequency catch up earlier. At the same time the interval for which the Fenlon solutions (4.101) and (4.122) are valid increases.

With increase in frequency of the high frequency, the amplitude of the high frequency coming from the harmonic of the low frequency is of course of smaller amplitude. Simultaneously, the number of oscillations of the high frequency fitting into the fixed interval increases, see the middle and lower curves in Figure 4.17.

In some applications it is desirable to nonlinearly create a low frequency from two higher frequencies, as the resulting sound beam is characterized by a narrow beamwidth and small sidelobes. This is called a parametric array and was first proposed by Westervelt (1960, 1963). The initial experimental verification (Berktay 1965) and further developments (Hobæk 1977; Naze Tjøtta and Tjøtta 1981; Zarembo and Krasilnikov 1979) were performed with water as medium. A book was published devoted entirely to the subject in Russia (Novikov, Rudenko and Timoshenko 1987). A disadvantage is that the energy conversion from high to low frequencies is poor. Still, parametric arrays are used in underwater applications, for example echo-location of bottom and sub-bottom. Design of parametric sources for underwater acoustics was discussed by Moffett and Mellen (1977) and Dybedal (1993).

In air the effect of nonlinearity is smaller than in water, but a parametric array in air was demonstrated by Bennett and Blackstock (1975). Yonegama and Fujimoto (1983) investigated the use of it for loudspeaker applications introducing the name *audio spotlight*. When using the parametric array for music, a particular difficulty is the nonlinear distortion which was discussed by Blackstock (1997^b).

The investigation of parametric array performances was made easier by the derivation of the KZK equation (see chapter 7) containing precisely the necessary parameters for the description of nonlinear sound beams in dissipative media .

Two sound waves colliding at nonzero angle also generate waves with sum and difference frequencies. This problem was studied since the 50's (Ingard and Pridmore-Brown 1956; Westervelt 1957; Berktay and Al-Temini 1969, 1971; Zverev and Kalachev 1970; Naugolnykh and Ostrovsky 1998). Colliding cylindrical waves from two separate sources were studied in the lossless case by Cary and Fenlon (1973).

Starting with initially two high frequencies in plane geometry - ignoring the beam diffraction - the difference frequency obtains part of the energy due to nonlinear interaction as seen in Figure 4.18. The higher frequencies are further nonlinearly damped more than the low through the shock movements. In reality, the viscous damping also attenuates higher frequencies more strongly than lower but in this chapter the viscous damping is zero.

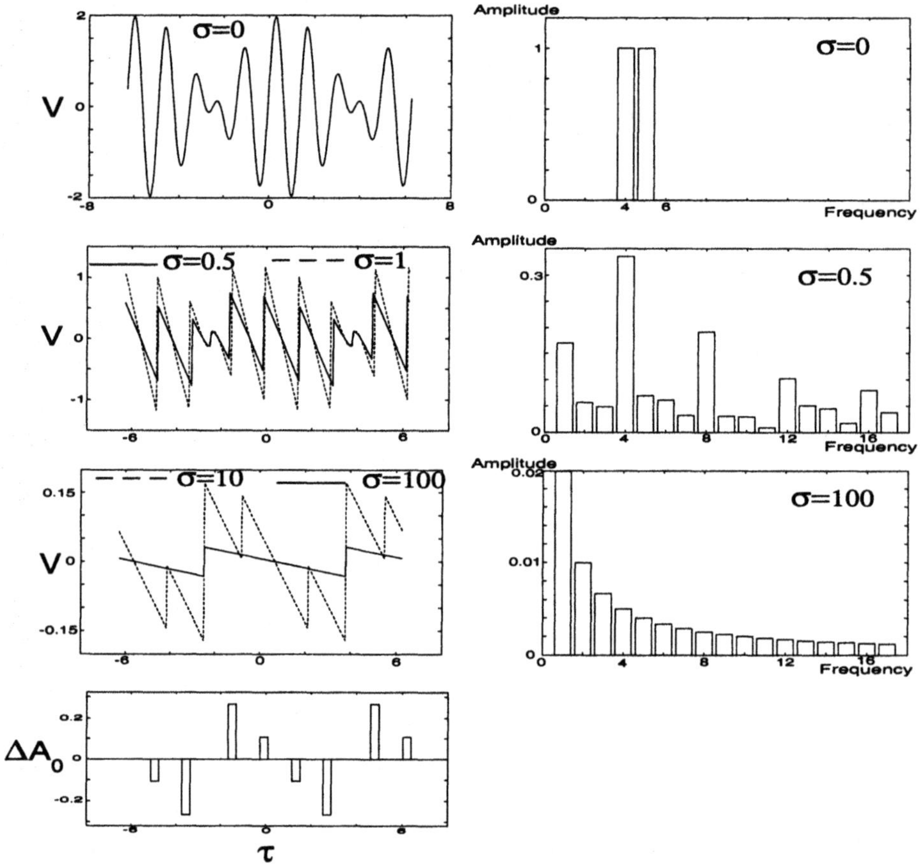

Figure 4.18: Parametric interaction creates a low frequency from two high frequencies. $V_0 = \sin(4\theta + \pi/5) + \sin(5\theta)$. The area difference between half periods, ΔA_0, was introduced in subsection 3.1.5.

The phases between frequencies are present in the Fenlon solution (4.101). It can be proved that phases have exactly this influence on all problems where the solution of a multifrequency condition can be expressed in the form of a multiple Fourier series (Hedberg 1996). A bifrequency source's final state (at long distances) has either one or two shocks per period depending on the choice of initial phase between the frequencies. Changing the initial phase results in change of amplitudes for the different frequency components of the propagated wave.

The phase may be determined through the phase theorem (Hedberg 1996) which states that when the solution to one set of initial phases whose solutions may be expressed as multiple Fourier series is known, the solution to all other sets of initial phases may be obtained. The amplitudes of frequencies are independent of initial phases when the ratios of the frequencies are irrational numbers. When the ratios are rational, they are dependent due to additions of components in the Fourier sum .

By using the phase theorem plus the Fenlon solution (4.101) for the boundary condition $V_0 = \sin(4\theta + \phi_1) + \sin(5\theta)$ the phase determined to yield maximum amplitude of the difference frequency $5 - 4 = 1$ is $\phi_1 = \pi/5$, see Figure 4.18.

The so-called degenerated case is defined by using the frequencies 1 and 2 and thus having the difference frequency equal one of the original frequencies. This is where the influence of phase is most notable. Let the original phase of frequency 2, ϕ_2, be allowed to vary in the boundary condition $V_0 = \sin\theta + \sin(2\theta + \phi_2)$. Through the phase theorem the optimal phase for maximum amplitude of frequency 2 is $\phi_2 = 0$. For maximum amplitude of frequency 1 the phase is $\phi_2 = \pi/2$ (Novikov and Rudenko 1976).

The results are seen in Figures 4.19 and 4.20. Note that also the amplitude of frequency two is slightly larger in Figure 4.20. The total energy content in Figure 4.19 is much smaller than in 4.20 due to the nonlinear extra attentuation taking place in the more pronounced shock.

Energy flowing upwards in the spectrum is natural through the nonlinear steepening of waves. We have also seen how energy is transferred downwards when difference frequencies are created from two higher frequencies. Now we will look at an example on how energy can flow to a lower frequency making its amplitude increase. Let in the degenerated case the boundary condition be $V_0 = \epsilon\sin\theta + \sin(2\theta + \phi_2)$ considering the limit of $\epsilon \to 0$. It can be shown how energy flows towards frequency 1 from frequency 2. At the distance of shock appearance the amplitude value is 1.18ϵ (Novikov and Rudenko 1976). The amplitude of frequency 1 will at large distances after shock formation increase towards twice its expected linear value as $\epsilon << 1$

(Kamakura, Ikegaya and Chou 1985).

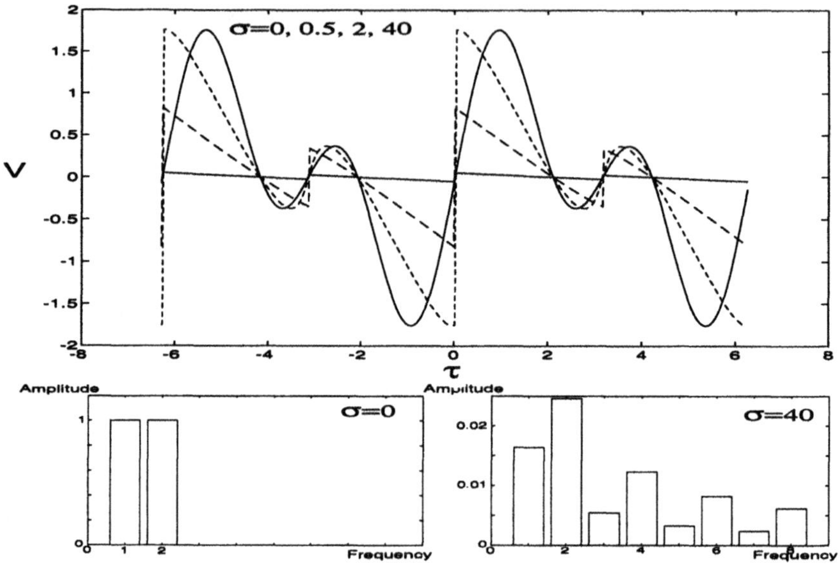

Figure 4.19: The first example of the degenerated case: $V_0 = \sin\theta + \sin 2\theta$

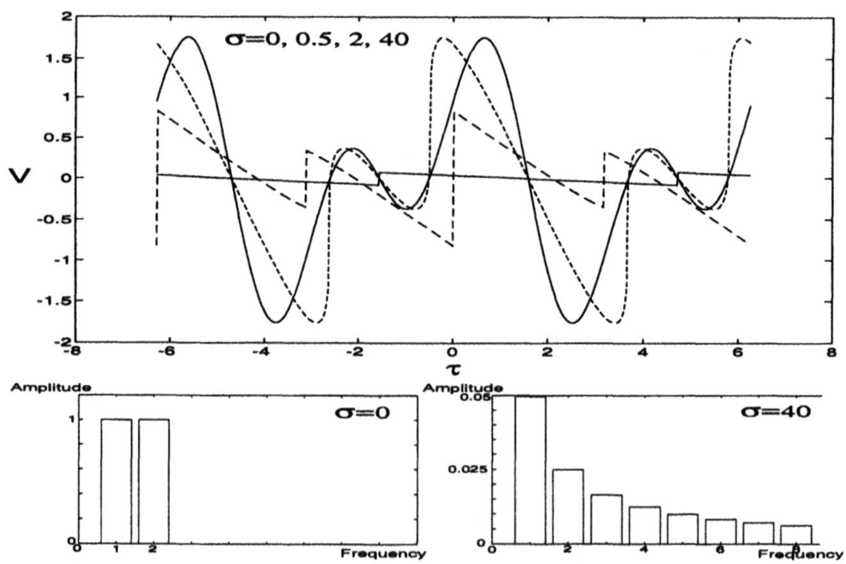

Figure 4.20: The second example of the degenerated case with different phase: $V_0 = \sin\theta + \sin(2\theta + \pi/2)$

The area differences between successive half-periods of the initial signal, first a negative and then the following positive, can be used to qualitatively estimate the long term behaviour of the wave evolution. The area difference is $\Delta A_0(\theta_n) = S_0(\theta_{n+1}) - S_0(\theta_n) = 0$ for all n where $S_0 = \int^\theta V(\theta', \sigma = 0)d\theta'$. If the area difference is zero the shock will not move on the retarded time axis θ, see Figure 4.19. If a shock is positive it will move to the left which means that the shock will arrive earlier in time, and in the opposite case (negative area difference) arrive later after having moved to the right, see Figure 4.20. When collisions of shocks take place the shocks merge and the area differences are added to estimate the merged shock's travelling state.

In the last rows in Figures 4.19 and 4.20 are showed the ΔA_0-plot (see page 45). In these are seen that none of the shocks moves for $\phi_2 = 0$, preserving the high frequency content, while two shocks per period move when $\phi_2 = \pi/2$ and the wave is transformed into a low frequency periodic sawtooth.

A description of an amplitude- and frequency-modulated wave is

$$V(\theta, \sigma = 0) = (1 + \epsilon \cos \Delta\theta)(\sin(b\theta + \frac{\vartheta b}{\Delta} \sin \Delta\theta)) \ , \qquad (4.123)$$

where b is the high frequency, Δ is the lower modulation frequency, ϵ is an amplitude modulation factor and ϑ is a frequency modulation factor. The behaviour is normally the expected with shocks forming, shocks moving and merging and the final result is a low frequency sawtooth wave, see Figure 4.21. (Gurbatov and Hedberg 1997). The following result is obtained by integration of (4.123):

$$\begin{aligned} S_0 &= \int^\theta V(\theta', \sigma = 0)d\theta' = \\ & -\frac{1 + \epsilon \cos(\Delta\theta)}{1 + \vartheta \cos(\Delta\theta)} \frac{1}{b} \cos(b\theta + \frac{\vartheta b}{\Delta} \sin(\Delta\theta)) + \\ & \int^\theta \frac{d}{d\theta'} [\frac{1 + \epsilon \cos \Delta\theta'}{1 + \vartheta \cos \Delta\theta'}] \frac{1}{b} \cos(b\theta' + \frac{\vartheta b}{\Delta} \sin \Delta\theta') d\theta' \end{aligned} \qquad (4.124)$$

The derivative $\frac{d}{d\theta'}[(1 + \epsilon \cos \Delta\theta')/(1 + \vartheta \cos \Delta\theta')]$ is small for $\Delta << b$ so that the integral term above is close to zero. It is exactly equal to zero when the amplitude modulation parameter is equal the frequency modulation parameter; $\epsilon = \vartheta$, for any Δ. This means the last integral in (4.124) is zero and the preceeding cosine-term has a constant amplitude. So the area differences between successive half-periods are equal to zero for all parts of the signal, i.e. $\Delta A_0(\theta_n) = S_0(\theta_{n+1}) - S_0(\theta_n) = 0$ for all n. To begin with the shocks move minutely, but will soon adapt to permanent positions. All shocks are still forever after the full shock structure has developed. This gives a constant shape of the wave form for all distances henceforth, only the amplitudes of the shocks are decreasing.

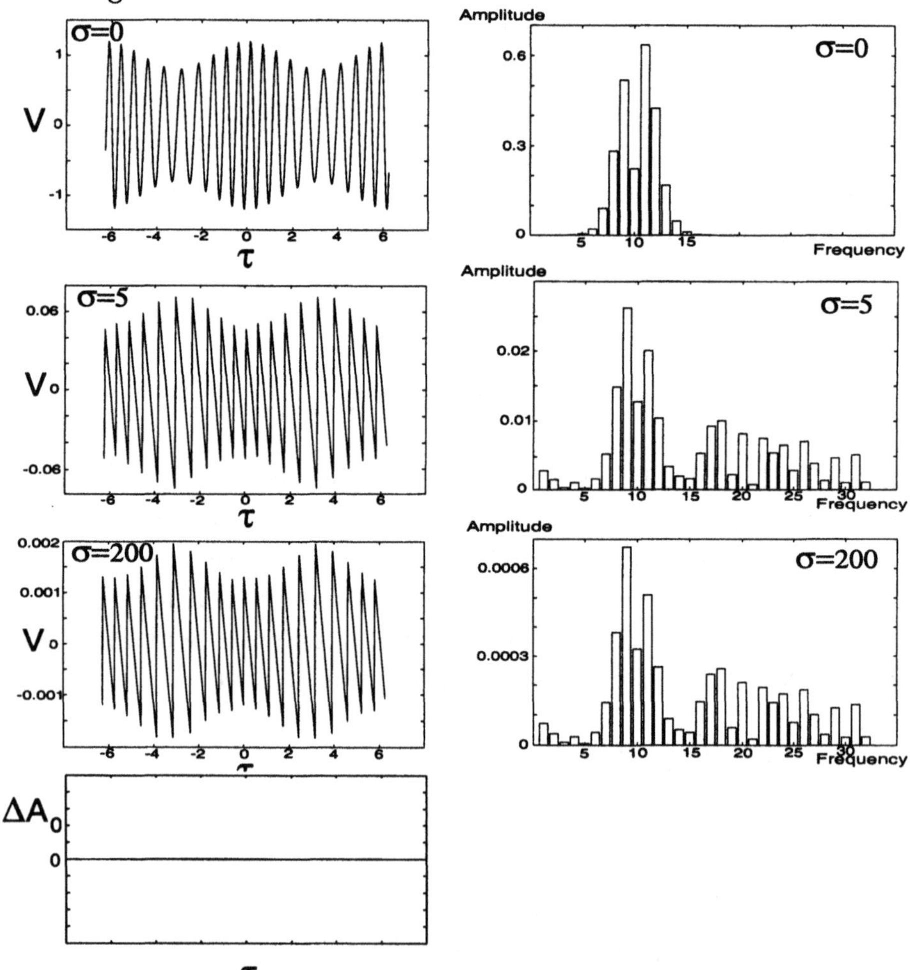

Figure 4.21: An amplitude- and frequency-modulated wave whose shocks are still for all distances.

Chapter 5

Nonlinear plane diffusive waves

The exact solution of Burgers' equation derived in section 3.2 will in this chapter be applied to different initial wave profiles in planar geometry. These refer to N-waves, harmonic waves, and multifrequency waves. For planar N-waves a method for finding the asymptotic wave will be given, which makes no use of the exact solution of Burgers' equation and thus prepares for treating nonplanar cases with no known exact solutions. For original planar harmonic waves two approximate solutions are derived in the shock region - the Fay solution and the Khokhlov-Soluyan solution. Both these solutions satisfy Burgers' equation, but they do not exactly satisfy the harmonic initial or boundary condition. The exact solution will be Fourier expanded in section 5.4 and joined with the corresponding Fourier expansion of the Riemann equation in the pre-shock region, that is, the Bessel-Fubini solution.

5.1 Planar N-waves

5.1.1 Shock solution

The importance of the N-wave lies in the fact that it is the final stage of an initial arbitrary short pulse with zero mean value propagating through a lossless (Gurbatov, Enflo and Pasmanik 1999) or slightly dissipative (Gurbatov, Enflo and Pasmanik 2001) medium. A study of the propagation of an arbitrary incident pulse of finite length without assuming small dissipation is made by Brander and Hedenfalk (1998).

An initial N-wave and its planar evolution is shown in Figure 5.1.

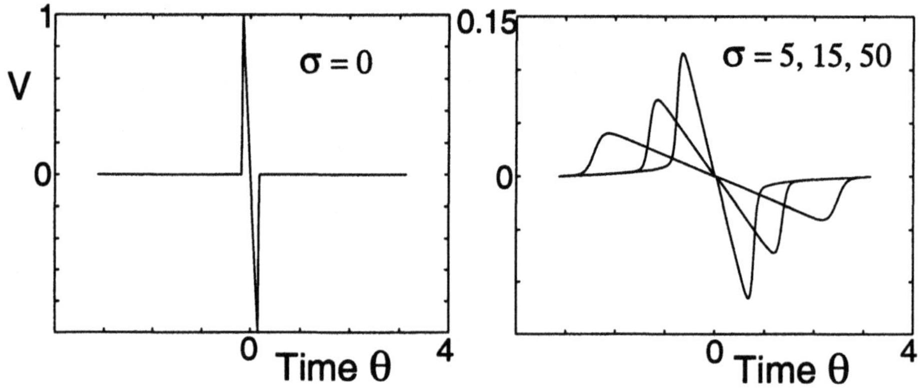

Figure 5.1: A planar N-wave.

As a first application of nonlinear wave propagation modelled by Burgers' equation we choose the N-wave studied as a solution of the inviscid wave equation (4.25). Burgers' equation from (2.79) is rewritten in the form

$$\frac{\partial v}{\partial t} + \beta v \frac{\partial v}{\partial \xi} - \frac{b}{2\rho_0}\frac{\partial^2 v}{\partial \xi^2} = 0 \ . \tag{5.1}$$

It is chosen as a model equation and is transformed into a dimensionless equation by the substitutions

$$V = \frac{v}{v_0} \tag{5.2}$$

$$T = 1 + \beta v_0 (t - t_0) \frac{1}{l_0} \tag{5.3}$$

$$X = \frac{\xi}{l_0} \ , \tag{5.4}$$

where l_0 is a characteristic length of the waveform in configuration space. Inserting (5.2)-(5.4) into (5.1) we obtain

$$\frac{\partial V}{\partial T} + V \frac{\partial V}{\partial X} = \epsilon \frac{\partial^2 V}{\partial X^2} \ , \tag{5.5}$$

where ϵ is related to the Reynolds number Re:

$$2\epsilon\beta = \frac{b}{\rho_0 l_0 v_0} \equiv \mathrm{Re}\ . \tag{5.6}$$

Note the difference in the definition of ϵ for N-waves and periodic waves (cf. (2.77)). We now assume the initial condition (4.28), (4.29) giving the solution $V_0(X,T)$ (4.26), (4.27) to the equation (4.25). The solution $V_0(X,T)$ to (4.25) can be considered as the first term in a series solution of equation (5.5):

$$V(X,T) = V_0(X,T) + \epsilon V_1(X,T) + \epsilon^2 V_2(X,T) + \dots \tag{5.7}$$

In this special case $V_0(X,T)$ solves (5.5) exactly, everywhere except for $X = T^{\frac{1}{2}}$, so that the expansion (5.7) can instead be written

$$V(X,T) = V_0(X,T) + o(\epsilon^n)\ , \tag{5.8}$$

where n is arbitrary. The expansion (5.7) is called an *outer* expansion, since it is valid in the region outside the neighbourhood of $X = T^{\frac{1}{2}}$ but is discontinuous at $X = T^{\frac{1}{2}}$.

In order to find an expression for the solution of (5.5) useful in the neighbourhood of the discontinuity of the solution $V_0(X,T)$ we make the change of variable

$$x^* = \frac{X - T^{\frac{1}{2}}}{\epsilon} \tag{5.9}$$

and rewrite (5.5) for the new dependent variable $V(x^*,T)$:

$$\epsilon\frac{\partial V}{\partial T} + (V - \frac{1}{2T^{\frac{1}{2}}})\frac{\partial V}{\partial x^*} = \frac{\partial^2 V}{\partial x^{*2}}\ , \tag{5.10}$$

A new *inner* expansion of $V(x^*,T)$ is made:

$$V(x^*,T) = V_0^*(x^*,T) + \epsilon V_1^*(x^*,T) + \epsilon^2 V_2^*(x^*,T) + \dots.. \tag{5.11}$$

Inserting (5.11) into (5.10) and integrating the equation for V_0^* gives

$$-\frac{V_0^*}{T^{\frac{1}{2}}} + V_0^{*2} - 2\frac{\partial V_0^*}{\partial x^*} = C(T)\ , \tag{5.12}$$

where $C(T)$ is an arbitrary function. Integration of (5.12) gives:

$$V_0^* = \frac{1}{2}T^{-\frac{1}{2}} - \left(\frac{C}{T^{\frac{1}{2}}} + \frac{1}{4T}\right)^{1/2} \tanh\left(\frac{x^* - A(T)}{2}\left(\frac{C}{T^{\frac{1}{2}}} + \frac{1}{4T}\right)^{\frac{1}{2}}\right) . \quad (5.13)$$

The solutions (5.13) and (5.8) now must join smoothly. This requires

$$\lim_{X \to T^{\frac{1}{2}}} V_0(X,T) = \lim_{x^* \to -\infty} V_0^*(x^*,T) . \quad (5.14)$$

The common limit for the expressions on both sides of (5.14) is attained for

$$T^{\frac{1}{2}} - X = O(\eta(\epsilon)) , \quad (5.15)$$

where $\eta(\epsilon)$ is chosen so that

$$\lim_{\epsilon \to 0} \eta = 0$$
$$\lim_{\epsilon \to 0} \frac{\eta}{\epsilon} = \infty . \quad (5.16)$$

The conclusion from (5.14) is that

$$C = 0 , \quad (5.17)$$

but that nothing can be said about $A(T)$. The equation (5.14) is an example of the *principle of asymptotic matching.* This principle is discussed more thoroughly in other books (Nayfeh 1973; Van Dyke 1975). An account for its use in nonlinear acoustics is given by Lesser and Crighton (1975).

In order to determine the function $A(T)$, which is a "correction due to diffusivity" of the position of the shock center we use the integral conservation technique (Crighton and Scott 1979). Integration of (5.5) from 0 to $X > T^{\frac{1}{2}}$ gives

$$\frac{d}{dT}\int_0^X V(x,T)\,dx$$
$$= -\epsilon\left(\left(\frac{\partial V}{\partial x}\right)_{x=0} - \left(\frac{\partial V}{\partial x}\right)_{x=X}\right) + \frac{1}{2}\left(V^2(0,T) - V^2(X,T)\right) . \quad (5.18)$$

For the righthand side of (5.18) we use the solution (4.26) and (4.27) and obtain

$$\frac{d}{dT}\int_0^X V(x,T)\,dx = -\frac{\epsilon}{T} . \quad (5.19)$$

Integration of (5.19) from $t = 1$ to $t = T$ gives after use of the initial condition (4.28) and (4.29),

$$\int_0^X V(x,T)\,dx = \frac{1}{2} - \epsilon \int_1^T \frac{1}{t}\,dt \qquad X > T^{\frac{1}{2}} \ . \tag{5.20}$$

For the evaluation of the integral on the lefthand side of (5.20) the outer solution $V_0(X,T)$ ((4.26) and (4.27)) and the inner solution V_0^* ((5.13) with $C = 0$) will be used. The first term of a uniform asymptotic expansion of V is obtained by adding V_0 and V_0^* and subtracting their common part V_c:

$$V = V_0 + V_0^* - V_c + O(\epsilon) \ , \tag{5.21}$$

where

$$V_0 = \frac{X}{T}(1 - H(X - T^{\frac{1}{2}})) \tag{5.22}$$

$$V_0^* = \frac{1}{2} T^{-\frac{1}{2}}(1 - \tanh \frac{\frac{x - T^{\frac{1}{2}}}{\epsilon} - A(T)}{4T^{\frac{1}{2}}}) \tag{5.23}$$

$$V_c = T^{-\frac{1}{2}}(1 - H(X - T^{\frac{1}{2}})) \ , \tag{5.24}$$

and $H(x)$ is the Heaviside function

$$\begin{aligned} H(x) &= 0 \ , \qquad & x < 0 \\ H(x) &= 1 \ , \qquad & x > 0 \ . \end{aligned} \tag{5.25}$$

Inserting (5.21)-(5.25) into the lefthand side of (5.20) and using the integral

$$\begin{aligned} &\int_0^{X > T^{\frac{1}{2}}} \tanh \frac{\frac{x - T^{\frac{1}{2}}}{\epsilon} - A(T)}{4T^{\frac{1}{2}}}\,dx \\ &= 4\epsilon T^{\frac{1}{2}}\Big(\ln\cosh \frac{X - T^{\frac{1}{2}} - \epsilon A(T)}{4\epsilon T^{\frac{1}{2}}} - \ln\cosh \frac{T^{\frac{1}{2}} + \epsilon A(T)}{4\epsilon T^{\frac{1}{2}}}\Big) \\ &\approx X - 2T^{\frac{1}{2}} - 2\epsilon A(T) \ , \end{aligned} \tag{5.26}$$

we obtain the result

$$A(T) = -T^{\frac{1}{2}} \int_1^T \frac{1}{t}\,dt \ , \tag{5.27}$$

which after integration, becomes

$$A(T) = -T^{\frac{1}{2}} \ln T \ . \tag{5.28}$$

It is easy to see that a term of $O(\epsilon)$ in (5.21) contributes to $O(\epsilon^2)$ on the lefthand side of (5.20) and thus does not affect (5.28). A shock wave like (5.23) was found by Taylor (1910) and is called a Taylor shock.

5.1.2 Old-age solution

We see from (5.23) and (5.28) that the shock center has moved from $X = T^{\frac{1}{2}}$ to $X = T^{\frac{1}{2}} - \epsilon T^{\frac{1}{2}} \ln T$ because of diffusivity. For T values of the order $T = O(e^{\frac{1}{2\epsilon}})$ this correction is of the same order of magnitude as the length of the N-wave. Then the asymptotic expansion (5.11) and the uniform lowest order solution (5.21) are no longer valid and the solution must be represented in some other way. In order to represent the solution for

$$T \geq T_1 \equiv \exp(\frac{1}{2\epsilon}) \tag{5.29}$$

we use the Cole-Hopf transformation (see (3.66) and (3.71)). The Cole-Hopf solution of (5.5) with the initial conditions (4.28) and (4.29) is (cf. (3.81))

$$V = -2\epsilon \frac{\partial U}{\partial X} / U \ , \tag{5.30}$$

where

$$U = \frac{1}{2\sqrt{\pi\epsilon(T-1)}} \int_{-\infty}^{\infty} U_0(x) \exp(-\frac{(X-x)^2}{4\epsilon(T-1)})\, dx \tag{5.31}$$

and

$$\begin{aligned} U_0(x) &= C \exp\Big(-\frac{1}{2\epsilon} \int_0^x V(\xi, 1)\, d\xi \Big) \\ &= \begin{cases} C \exp(-x^2/(4\epsilon)) & x^2 < 1 \\ C \exp(-1/4\epsilon) & x^2 > 1 \ . \end{cases} \end{aligned} \tag{5.32}$$

If we suppress an unimportant multiplicative constant, the evaluation of the integral (5.31) gives,

$$U(X,T) = \exp(-\frac{X^2}{4\epsilon T})\frac{1}{2T^{1/2}}$$
$$\Big\{\mathrm{erf}[\frac{1}{2}(\frac{T}{\epsilon(T-1)})^{1/2}(1+\frac{X}{T})] + \mathrm{erf}[\frac{1}{2}(\frac{T}{\epsilon(T-1)})^{1/2}(1-\frac{X}{T})]\Big\}$$
$$+\frac{1}{2}\exp(-\frac{1}{4\epsilon T})\Big\{2+\mathrm{erf}(\frac{X-1}{2(\epsilon(T-1))^{1/2}}) - \mathrm{erf}(\frac{X+1}{2(\epsilon(T-1))^{1/2}})\Big\} . \quad (5.33)$$

The exact expression (5.33) can be approximated for large values of T. We therefore assume:

$$1 \ll X \ll T = O(\mathrm{e}^{1/(2\epsilon)}) . \quad (5.34)$$

Using the formula for the error function for large arguments,

$$\mathrm{erf}(\sqrt{x}) \approx 1 - \frac{1}{\sqrt{\pi}}\frac{\mathrm{e}^{-x}}{\sqrt{x}} , \quad (5.35)$$

and the derivative of the error function

$$\frac{d(\mathrm{erf}\, x)}{dx} = \frac{2}{\sqrt{\pi}}\mathrm{e}^{-x^2} , \quad (5.36)$$

we find the leading contributions to $U(X,T)$:

$$U(X,T) \approx 2\epsilon^{\frac{1}{2}}\sqrt{\pi}\Big(\exp(-\frac{X^2}{4\epsilon T}) + T^{\frac{1}{2}}\exp(-\frac{1}{4\epsilon})\Big) . \quad (5.37)$$

We introduce the scaled variables X', T' and V' according to the substitutions

$$\begin{aligned} X' &= \epsilon^{-\frac{1}{2}}T_1^{-\frac{1}{2}}X \\ T' &= T_1^{-1}T \\ V' &= \epsilon^{-\frac{1}{2}}T_1^{\frac{1}{2}}V \end{aligned} \quad (5.38)$$

and obtain from the result (5.37) and (5.30)

$$V' = \frac{X'}{T'}\Big(T'^{\frac{1}{2}}\exp(\frac{X'^2}{4T'}) + 1\Big)^{-1} . \quad (5.39)$$

It is remarkable that the solution (5.39) is an exact solution of Burgers' equation

$$\frac{\partial V'}{\partial T'} + V'\frac{\partial V'}{\partial X'} = \frac{\partial^2 V'}{\partial X'^2}, \tag{5.40}$$

obtained from (5.5) using (5.38).

For $T' = O(1)$ both terms in the denominator of the solution (5.39) are of the same order of magnitude, implying that the nonlinear term in (5.40) cannot be neglected. For $T' \gg 1$ (the old-age region) the solution (5.39) can be approximated:

$$V' \approx \frac{X'}{T'^{3/2}} \exp\left(-\frac{X'^2}{4T'}\right) \tag{5.41}$$

The solution (5.41), called the old-age solution, is an exact solution of the linear equation

$$\frac{\partial V'}{\partial T'} = \frac{\partial^2 V'}{\partial X'^2}. \tag{5.42}$$

5.1.3 The old-age solution found by an alternative method

By means of the Cole-Hopf solution (5.30) we have obtained the solution of the (linear) old-age equation (5.42), which develops with growing T from the solution (5.22) and (5.23) (called the Taylor shock solution) of the nonlinear equation (5.5). Both the waveform and the amplitude constant (in this case unity) are uniquely determined in the representation (5.41). However, the success is dependent on the existence of the exact solution representation of (5.5). For most physically interesting nonlinear wave equations exact solution are not known. Examples are generalized Burgers' equations which describe nonlinear acoustic waves in spherical and cylindrical geometry; they are treated later in this book. A new technique has been developed (Enflo 1998) to find the old-age waveform and its amplitude constant for such cases. This technique will now be illustrated with the help of the problem just treated. We will see how the old-age solution (5.41) can be found without reference to the Cole-Hopf transformation.

The procedure for finding the old-age solution developing from a shock solution can be divided into four steps.

Step 1) From Taylor's shock solution (in the present case (5.23)) a solution is found valid in a region where X is greater than $T^{\frac{1}{2}}$, $T = O(1)$ and the nonlinear term in (5.5) is neglected. This region is called the fading shock (or shock tail) region.

Step 2) After rescaling of the original equation without the nonlinear term the linear equation valid in the old-age regime is found (in the present case (5.42)); the general solution of this equation is written as an integral containing an unknown function.

Step 3) The integral representation of the solution of the linear equation mentioned in Step 2 is evaluated in the fading shock region by means of the steepest descent (or saddle-point) method. By identification with the fading shock solution found in Step 1 the unknown function in the integral representation is determined.

Step 4) The integral representation found is evaluated in order to find its leading contribution in the old-age regime. In the present case of plane waves this evaluation can be done exactly.

Now the four steps above will be executed in detail for the problem of finding the old-age behaviour of a plane N-wave.

1) Solution in the fading shock (or shock tail) region

According to the equations (5.23) and (5.28) Taylor's shock solution is

$$V_0^* = \frac{1}{2}T^{-\frac{1}{2}}\left(1 - \tanh\frac{X - T^{\frac{1}{2}} + \epsilon T^{\frac{1}{2}}\ln T}{4\epsilon T^{\frac{1}{2}}}\right) . \tag{5.43}$$

After the change of variable

$$Y = \frac{X - T^{\frac{1}{2}}}{2\epsilon T^{\frac{1}{2}}} , \tag{5.44}$$

(5.30) becomes

$$V_0^* = \frac{T^{-\frac{1}{2}}}{1 + T^{\frac{1}{2}} \exp(Y)} \,. \tag{5.45}$$

Now Y must be chosen so that V_0^* becomes small, thus $V = O(\delta)$ with $\delta \ll 1$. This happens when

$$Y > \ln \frac{1}{\delta} \,, \tag{5.46}$$

while

$$T = O(1) \,. \tag{5.47}$$

Equations (5.46) and (5.47) define the fading shock region where the lowest order solution in δ is

$$V_0^* = T^{-1} \exp(-Y) \,. \tag{5.48}$$

It is easy to check that the nonlinear term in (5.5) is of lower order of magnitude in δ than the other two terms if the solution (5.48) along with (5.46) and (5.47) is inserted therein.

2) Integral representation of the solution in the old-age region

A general solution of the old-age equation (5.42) is

$$V' = \int_{-\infty}^{\infty} \exp(-\lambda^2 T' + i\lambda X') h(\lambda) \, d\lambda \,. \tag{5.49}$$

The function $h(\lambda)$ must be determined using the information in Step 1.

3) Evaluation of the old-age integral representation in the fading shock region

The integral (5.49) is evaluated in the fading shock region. To that end the variables (x', T') are changed to (Y, T). In order that the two terms of the argument of the exponential function in (5.35) balance each other, the variable of integration is changed from λ to κ:

$$\lambda = \epsilon^{-\frac{1}{2}} \exp(\frac{1}{4\epsilon})\kappa \tag{5.50}$$

Putting

$$h(\lambda) = \exp(\frac{f(\kappa)}{\epsilon})g(\kappa) \tag{5.51}$$

equation (5.49) becomes

$$V' = \epsilon^{-\frac{1}{2}} \exp(\frac{1}{4\epsilon}) \int_{-\infty}^{\infty} \exp[-\frac{1}{\epsilon}(\kappa^2 T - i\kappa T^{\frac{1}{2}} - f(\kappa)) + 2i\kappa Y T^{\frac{1}{2}}] g(\kappa)\, d\kappa \;. \tag{5.52}$$

The result of the evaluation of the integral (5.52) will be identified with the fading shock solution (5.48). Putting

$$\chi(\kappa) = -\kappa^2 T + i\kappa T^{\frac{1}{2}} + f(\kappa) \;, \tag{5.53}$$

this identification gives

$$\chi(\kappa_0) = 0 \tag{5.54}$$

$$\kappa_0 = \tfrac{i}{2} T^{-\frac{1}{2}} \quad , \tag{5.55}$$

where κ_0 is the saddle-point given by

$$\chi'(\kappa_0) = 0 \;. \tag{5.56}$$

From (5.53)-(5.55) we find

$$f(\kappa_0) = \frac{1}{4} \;. \tag{5.57}$$

The integral (5.52) by the steepest descent method is found to be

$$V' = \sqrt{\pi} \exp(\frac{1}{4\epsilon}) \exp(-Y) T^{-\frac{1}{2}} g(\frac{i}{2} T^{-\frac{1}{2}}) \;. \tag{5.58}$$

Comparing (5.48) and (5.58), and using (5.38) and (5.11) we find that

$$g(\kappa) = -2i\epsilon^{-\frac{1}{2}} \pi^{\frac{1}{2}} \kappa \;. \tag{5.59}$$

Inserting (5.57) and (5.59) into (5.52) and changing the integration variable according to (5.50) a representation of the old-age wave function is obtained, with the unknown function $h(\lambda)$ in (5.49) fully determined:

$$V'(X',T') = -2i\pi^{-\frac{1}{2}} \int_{-\infty}^{\infty} \exp(-\lambda^2 T' + i\lambda X')\lambda \, d\lambda \; . \qquad (5.60)$$

This expression gives the lowest-order solution in ϵ of the problem of determining the old-age wave evolving from the shock solution (5.21).

4) Evaluation of the old-age integral representation

In the present case the integral representation (5.60) of the old-age wave can be evaluated exactly, by using the integral representation of Hermite functions (Lebedev (1965), chapter 10):

$$H_\nu(z) = \frac{2^\nu \exp(z^2)}{\pi^{\frac{1}{2}}} \int_0^\infty \exp(u^2 + 2izu - \frac{1}{2}\nu\pi)u^\nu \, du + c.c. \qquad (5.61)$$

Here, c.c. denotes the complex conjugate. From (5.61) it is obviuos that the Hermite polynomial $H_1(\frac{X'}{2\sqrt{T'}}) = \frac{X'}{\sqrt{T'}}$ is obtained from the righthand side of (5.60):

$$V'(X',T') = T'^{-1} \exp(-\frac{X'^2}{4T'}) H_1(\frac{X'}{2\sqrt{T'}}) = T'^{-\frac{3}{2}} X' \exp(-\frac{X'^2}{4T'}) \; . \qquad (5.62)$$

Thus (5.62) gives the same old-age behaviour as (5.41), derived here without reference to the Cole-Hopf transformation. The method used in deriving (5.62) will be employed later to obtain the old-age behaviour of N-waves for non-planar geometries.

5.2 Planar harmonic waves. The Fay solution

5.2.1 Derivation of Fay's solution from the Cole-Hopf solution

As an example of the evolution of a planar harmonic wave we study a sine wave generated by a vibrating piston at the end of an infinite pipe with constant transverse area and filled with a fluid. The time dependent boundary condition at the end of the pipe, i.e. at $x = 0$ is

$$v = v_0 \sin \omega t \ , \tag{5.63}$$

where v is the fluid velocity. In dimensionless variables, our problem is to solve Burgers' equation (2.76) with the boundary condition

$$V(\sigma = 0, \theta) = \sin \theta \ . \tag{5.64}$$

The solution of (2.76) is now given by the Cole-Hopf formula (3.71) and (3.83). The Fourier coefficients, A_k, in (3.83) are given by (3.87) and (3.88) with $V_0(\theta) = \sin \theta$. Thus

$$A_0 = \frac{C}{\pi} \int_0^\pi \exp\Big(\frac{1}{2\epsilon}(1 - \cos\theta)\Big)\, d\theta \tag{5.65}$$

$$A_k = \frac{2C}{\pi} \int_0^\pi \exp\Big(\frac{1}{2\epsilon}(1 - \cos\theta)\Big) \cos k\theta \, d\theta \ . \tag{5.66}$$

Using the integral representation of the modified Bessel function of integer order $I_{\pm n}(z)$

$$I_{\pm n}(z) = \frac{1}{\pi} \int_0^\pi \mathrm{e}^{z \cos\theta} \cos n\theta \, d\theta \ , \tag{5.67}$$

we obtain from (5.65) and (5.66),

$$A_0 = C \mathrm{e}^{\frac{1}{2\epsilon}} I_0\Big(-\frac{1}{2\epsilon}\Big) \ , \tag{5.68}$$

$$A_k = 2C \mathrm{e}^{\frac{1}{2\epsilon}} I_k\Big(-\frac{1}{2\epsilon}\Big) \ . \tag{5.69}$$

Using (3.71) and the formula

$$I_n(-z) = (-1)^n I_n(z) \ , \tag{5.70}$$

the solution to (2.76) evolving from the harmonic boundary condition (5.64) is found to be (Cole 1951; Mendousse 1953)

$$V(\sigma,\theta) = -4\epsilon \frac{\sum_{n=1}^{\infty} \exp(-n^2\epsilon\sigma) n(-1)^n I_n(\frac{1}{2\epsilon}) \sin n\theta}{I_0(\frac{1}{2\epsilon}) + 2\sum_{n=1}^{\infty} \exp(-n^2\epsilon\sigma)(-1)^n I_n(\frac{1}{2\epsilon}) \cos n\theta} \ . \tag{5.71}$$

It is seen from the formula (see Gradshteyn and Ryzhik (1965), 8.514)

$$\exp(z\cos\theta) = I_0(-z) + 2\sum_{n=1}^{\infty} I_n(-z)\cos n\theta \tag{5.72}$$

that the solution (5.71) satisfies the boundary condition (5.64).

We will now show how the exact solution (5.71) can be approximated by a single Fourier series. Assume that this series is

$$V(\sigma,\theta) = \sum_{n=1}^{\infty} a_n(\sigma) \sin n\theta \ . \tag{5.73}$$

Put

$$b_n(\sigma) = \exp(-n^2\epsilon\sigma)(-1)^n I_n(\frac{1}{2\epsilon}) \ . \tag{5.74}$$

Because $I_n(z)$ is even in n we define, generalizing (5.74):

$$b_{-n}(\sigma) = b_n(\sigma) \ . \tag{5.75}$$

Equating (5.73) and (5.71) we have

$$\sum_{k=1}^{\infty} a_k \sin k\theta \Big[b_0 + 2\sum_{l=1}^{\infty} b_l \cos l\theta \Big] = -4\epsilon \sum_{m=1}^{\infty} m b_m \sin m\theta \ . \tag{5.76}$$

The lefthand side of (5.76) is rewritten as

$$\sum_{k=1}^{\infty} a_k \sin k\theta \Big[b_0 + 2\sum_{l=1}^{\infty} b_l \cos l\theta \Big]$$

$$= \sum_{m=1}^{\infty} b_0 a_m \sin m\theta + \sum_{k=1}^{\infty}\sum_{l=1}^{\infty} a_k b_l\Big(\sin(k+l)\theta + \sin(k-l)\theta\Big)$$

$$= \sum_{m=1}^{\infty} a_m b_0 \sin m\theta + \sum_{m=2}^{\infty}\sum_{k=1}^{m-1} a_k b_{m-k} \sin m\theta$$

$$+ \sum_{m=1}^{\infty}\sum_{k=m+1}^{\infty} a_k b_{k-m} \sin m\theta - \sum_{m=1}^{\infty}\sum_{l=m+1}^{\infty} a_{l-m} b_l \sin m\theta$$

$$= \sum_{m=1}^{\infty} \sin m\theta \Big[a_m b_0 + \sum_{k=1,k\neq m}^{\infty} a_k b_{k-m} - \sum_{k=1}^{\infty} a_k b_{k+m}\Big]$$

$$= \sum_{m=1}^{\infty} \sin m\theta \sum_{k=1}^{\infty} a_k (b_{k-m} - b_{k+m}) \; . \tag{5.77}$$

Inserting (5.77) into (5.76) and using (5.74) we obtain a linear system of equations for the Fourier coefficients $a_n(\sigma)$ in the expansion (5.73):

$$\sum_{k=1}^{\infty} a_k (-1)^k \mathrm{e}^{-k^2\epsilon\sigma}\Big(\mathrm{e}^{2km\epsilon\sigma} I_{k-m}(\frac{1}{2\epsilon}) - \mathrm{e}^{-2km\epsilon\sigma} I_{k+m}(\frac{1}{2\epsilon})\Big) = -4\epsilon m I_m(\frac{1}{2\epsilon}) \; . \tag{5.78}$$

The system of equations (5.78) was given by Blackstock (1964). The exact solution of this system can be found through a recursion formula for the coefficients in a power series expansion of a_k, as we will see later. First, however, we shall study some approximate representations of the periodic solutions of Burgers' equation (2.76) with the boundary condition (5.64).

In order to obtain an approximate representation of $V(\sigma,\theta)$, exactly given by (5.71) or by (5.73) with (5.78) we will assume that ϵ is a small number so that asymptotic representations of the modified Bessel functions $I_n(\frac{1}{2\epsilon})$ for large arguments can be used. The representation to be used is (see Gradshteyn and Ryzhik 1965, 8.451.5)

$$I_n(\frac{1}{2\epsilon}) = (\frac{\epsilon}{\pi})^{\frac{1}{2}} \exp(\frac{1}{2\epsilon})\Big(1 - \epsilon\frac{4n^2-1}{4} + O(\epsilon^2)\Big) \; . \tag{5.79}$$

Using only the first term of (5.79) and writing

$$q = \mathrm{e}^{-\epsilon\sigma} \; , \tag{5.80}$$

we obtain from (5.71) in approximation

$$V(\sigma,\theta) \approx -4\epsilon \frac{\sum_{n=1}^{\infty} n(-1)^n q^{n^2} \sin n\theta}{1 + 2\sum_{n=1}^{\infty} q^{n^2}(-1)^n \cos n\theta} \,, . \tag{5.81}$$

Expression (5.81) is valid according to (5.79) only if terms with $n \sim \epsilon^{-\frac{1}{2}}$ give small contributions to the sums in the nominator and denominator. This is the case for $\sigma \gg 1$. For $\sigma \approx 4$ the relative error in the series (5.81) in comparison with the exact solution (5.71) is of order 10^{-2}.

The solution (5.81) can be written as a Fourier series. This can be seen by use of ϑ(theta)-functions but also directly by elementary methods. We first give the ϑ-function argument (Cole 1951). The function $\vartheta_3(z,q)$ is defined as

$$\vartheta_3(z,q) = 1 + 2\sum_{n=1}^{\infty} q^{n^2} \cos 2nz \ . \tag{5.82}$$

The denominator in (5.81) thus is $\vartheta_3(\frac{\theta}{2}, -\mathrm{e}^{-\epsilon\sigma})$. For $\vartheta_3(z,q)$ the following identity is valid (Whittaker and Watson 1950):

$$\frac{\partial\vartheta_3}{\partial z}\frac{1}{\vartheta_3} = 4\sum_{n=1}^{\infty} \frac{(-1)^n q^n \sin 2nz}{1-q^{2n}} = 4\sum_{n=1}^{\infty} \frac{(-1)^{n+1} \sin 2nz}{q^n - q^{-n}} \ . \tag{5.83}$$

From (5.81) and (5.82) we find

$$V(\sigma,\theta) = 2\epsilon \frac{\partial\vartheta_3(\frac{\theta}{2}, -\mathrm{e}^{-\epsilon\sigma})}{\partial\theta} \frac{1}{\vartheta_3(\frac{\theta}{2}, -\mathrm{e}^{-\epsilon\sigma})} \ . \tag{5.84}$$

Comparison of (5.83) and (5.84) gives

$$V(\sigma,\theta) = 4\epsilon \sum_{n=1}^{\infty} \frac{(-1)^{n+1} \sin n\theta}{(-1)^n(\mathrm{e}^{-n\epsilon\sigma} - \mathrm{e}^{n\epsilon\sigma})} = 2\epsilon \sum_{n=1}^{\infty} \frac{\sin n\theta}{\sinh n\epsilon\sigma} \ . \tag{5.85}$$

The solution (5.85) was found by R.D. Fay (1931). Although the approximation (5.79) has been made in the derivation, (5.85) is an exact solution of Burgers' equation (2.76) as we shall prove presently. Observe that this is an approximate solution (not exact) for the sine wave *boundary condition.*

5.2.2 Direct derivation of Fay's solution

We will now give a direct derivation of Fay's solution (5.85). For this derivation the following identity is needed:

$$\sum_{m=1}^{\infty}(-1)^{m+1}q^{m^2}(q^{(2l-1)m}+q^{-(2l-1)m})=1\ , \tag{5.86}$$

where q is arbitrary, $|q|<1$, and l is an arbitrary integer. The proof of (5.86) consists of a straightforward summation of the left-hand side:

$$\begin{aligned}
&\sum_{m=1}^{\infty}(-1)^{m+1}q^{m^2}(q^{(2l-1)m}+q^{-(2l-1)m})\\
&=-\sum_{m=1}^{\infty}(-1)^{m}q^{m(m+2l-1)}-\sum_{m=-2l+2}^{\infty}(-1)^{m+2l-1}q^{m(m+2l-1)}\\
&=-\sum_{m=1}^{\infty}(-1)^{m}q^{m(m+2l-1)}-\sum_{m=-2l+2}^{0}(-1)^{m+1}q^{m(m+2l-1)}\\
&\quad+\sum_{m=1}^{\infty}(-1)^{m}q^{m(m+2l-1)}=\sum_{m=0}^{2l-2}(-1)^{m}q^{m(m-2l+1)}\\
&=1-q^{-2l+2}+q^{2(-2l+3)}-q^{3(-2l+4)}+\dots\\
&\quad\dots-q^{(2l-3)(2l-3-2l+1)}+q^{(2l-2)(2l-2-2l+1)}\\
&=1\ .
\end{aligned} \tag{5.87}$$

Since (5.86) is valid for arbitrary integers l, we give l the values $1,2,3,.....k$ and sum over l in (5.86):

$$\sum_{m=1}^{\infty}(-1)^{m+1}q^{m^2}\sum_{l=1}^{k}(q^{(2l-1)m}+q^{-(2l-1)m})=k\ . \tag{5.88}$$

Because of the identity

$$\sum_{l=1}^{k}(q^{(2l-1)m}+q^{-(2l-1)m})=\frac{(q^m)^{2k}-(q^{-m})^{2k}}{q^m-q^{-m}}\ , \tag{5.89}$$

equation (5.88) can be written as

$$k=\sum_{m=1}^{\infty}(-1)^m\frac{q^m}{1-q^{2m}}(q^{m^2+2mk}-q^{m^2-2mk})\ . \tag{5.90}$$

Making the approximation that all modified Bessel functions $I_n(\frac{1}{2\epsilon})$ are equal, which means that only the first term is retained on the righthand side of (5.79), and using (5.80) we rewrite (5.78) as

$$k = \frac{1}{4\epsilon} \sum_{m=1}^{\infty} a_m (-1)^m q^{m^2} (q^{2mk} - q^{-2mk}) . \tag{5.91}$$

A comparison of (5.90) and (5.91) gives the Fourier coefficient a_m:

$$a_m = 4\epsilon \frac{e^{-m\epsilon\sigma}}{1 - e^{-2m\epsilon\sigma}} = 2\epsilon \frac{1}{\sinh m\epsilon\sigma} , \tag{5.92}$$

in agreement with (5.85).

5.2.3 Proof that Fay's solution satisfies Burgers' equation

We will now prove that Fay's solution (5.85) is an exact solution of Burgers' equation (2.76). We call Fay's solution V^f and rewrite (5.85) as

$$V^f(\sigma,\theta) = 2\epsilon \sum_{n=1}^{\infty} \frac{\sin n\theta}{\sinh n\epsilon\sigma} = \frac{2\epsilon}{i} \sum_{n=1}^{\infty} (e^{in\theta} - e^{-in\theta}) \frac{e^{-n\epsilon\sigma}}{1 - e^{-2n\epsilon\sigma}} . \tag{5.93}$$

From (5.93) we find that

$$\begin{aligned}
&\frac{\partial V^f}{\partial \sigma} - V^f \frac{\partial V^f}{\partial \theta} - \epsilon \frac{\partial^2 V^f}{\partial \theta^2} \\
&\quad = \frac{2\epsilon^2}{i} \Big[-\sum_{n=1}^{\infty} (e^{in\theta} - e^{-in\theta}) n \frac{e^{-n\epsilon\sigma}(1 + e^{-2n\epsilon\sigma})}{(1 - e^{-2n\epsilon\sigma})^2} \\
&\qquad - 2 \sum_{n=1}^{\infty} \frac{(e^{in\theta} - e^{-in\theta}) e^{-n\epsilon\sigma}}{(1 - e^{-2n\epsilon\sigma})} \sum_{m=1}^{\infty} m (e^{im\theta} + e^{-im\theta}) \frac{e^{-m\epsilon\sigma}}{1 - e^{-2m\epsilon\sigma}} \\
&\qquad + \sum_{n=1}^{\infty} n^2 (e^{in\theta} - e^{-in\theta}) \frac{e^{-n\epsilon\sigma}}{1 - e^{-2n\epsilon\sigma}} \Big] = \frac{2\epsilon^2}{i} \sum_{p=-\infty}^{\infty} A(p) e^{ip\theta} .
\end{aligned} \tag{5.94}$$

The proof is complete if we can show that $A(p) = 0$. From (5.94) we have, for $p > 0$,

$$A(p) = -p \frac{e^{-p\epsilon\sigma}(1 + e^{-2p\epsilon\sigma})}{(1 - e^{-2p\epsilon\sigma})^2}$$

$$
\begin{aligned}
&-2\sum_{n=1}^{p-1}(p-n)\frac{e^{-n\epsilon\sigma}}{1-e^{-2n\epsilon\sigma}}\frac{e^{-(p-n)\epsilon\sigma}}{1-e^{-2(p-n)\epsilon\sigma}}\\
&-2\sum_{n=p+1}^{\infty}(n-p)\frac{e^{-n\epsilon\sigma}}{1-e^{-2n\epsilon\sigma}}\frac{e^{-(n-p)\epsilon\sigma}}{1-e^{-2(n-p)\epsilon\sigma}}\\
&+2\sum_{n=p+1}^{\infty}n\frac{e^{-n\epsilon\sigma}}{1-e^{-2n\epsilon\sigma}}\frac{e^{-(n-p)\epsilon\sigma}}{1-e^{-2(n-p)\epsilon\sigma}}+p^2\frac{e^{-p\epsilon\sigma}}{1-e^{-2p\epsilon\sigma}}\,. \qquad (5.95)
\end{aligned}
$$

Adding the third and the fourth terms on the right-hand side of (5.95) and putting $e^{-2\epsilon\sigma}=x$ we obtain

$$
\begin{aligned}
&A(p)x^{-\frac{1}{2}p}\\
&=-p\frac{1+x^p}{(1-x^p)^2}+p^2\frac{1}{1-x^p}-2p\sum_{n=1}^{p-1}\frac{1}{1-x^n}\frac{1}{1-x^{p-n}}\\
&+2\sum_{n=1}^{p-1}n\frac{1}{1-x^n}\frac{1}{1-x^{p-n}}+2p\sum_{n=p+1}^{\infty}\frac{x^{n-p}}{(1-x^n)(1-x^{n-p})}\,. \qquad (5.96)
\end{aligned}
$$

Using the identity

$$
\begin{aligned}
2\sum_{n=1}^{p-1}(p-n)\frac{1}{1-x^n}\frac{1}{1-x^{p-n}}&=2\sum_{n=1}^{p-1}n\frac{1}{1-x^n}\frac{1}{1-x^{p-n}}\\
&=\sum_{n=1}^{p-1}p\frac{1}{1-x^n}\frac{1}{1-x^{p-n}}\,, \qquad (5.97)
\end{aligned}
$$

we rewrite (5.96) as

$$
\begin{aligned}
&A(p)x^{-\frac{1}{2}p}\\
&=-p\frac{1+x^p}{(1-x^p)^2}+p^2\frac{1}{1-x^p}\\
&-\sum_{n=1}^{p-1}p\frac{1}{1-x^n}\frac{1}{1-x^{p-n}}+2p\sum_{n=p+1}^{\infty}\frac{x^{n-p}}{(1-x^n)(1-x^{n-p})}\\
&=-p\frac{1+x^p}{(1-x^p)^2}+p^2\frac{1}{1-x^p}\\
&-\frac{p}{1-x^p}\sum_{n=1}^{p-1}\left(\frac{1}{1-x^n}+\frac{x^{p-n}}{1-x^{p-n}}\right)+\frac{2p}{1-x^p}\sum_{n=1}^{\infty}\left(\frac{x^n}{1-x^n}-\frac{x^{p+n}}{1-x^{p+n}}\right)
\end{aligned}
$$

$$= \frac{p}{1-x^p}\Big[-\frac{1+x^p}{1-x^p}+p-\sum_{n=1}^{p-1}\frac{1}{1-x^n}-\sum_{n=1}^{p-1}\frac{x^n}{1-x^n}+2\sum_{n=1}^{p}\frac{x^n}{1-x^n}\Big]$$

$$= \frac{p}{1-x^p}\Big[-\frac{1+x^p}{1-x^p}+p+2\frac{x^p}{1-x^p}-\sum_{n=1}^{p-1}\frac{1-x^n}{1-x^n}\Big]$$

$$= \frac{p}{1-x^p}\Big[\frac{1}{1-x^p}(-1-x^p+2x^p)+p-(p-1)\Big]=0\ . \tag{5.98}$$

It is obvious that the proof that $A(p)=0$ is also valid for $p<0$. We have thus shown that Fay's solution V^f, given by (5.93), is an exact solution of Burgers' equation.

5.2.4 Some notes on Fay's solution

Fay's solution can be refined by taking the second term in the asymptotic expansion (5.79) for use in (5.78). Because the essential information in the approximate system of equations (5.91) is contained in the terms where

$$\epsilon(k\pm m)^2 \ll 1 \tag{5.99}$$

we can approximate (5.79) by writing ("$\sim$" means "proportional to")

$$I_{k\pm m}(\frac{1}{2\epsilon}) \sim 1-\epsilon(k\pm m)^2+\frac{\epsilon}{4} \approx \exp\big(-\epsilon(k\pm m)^2+\frac{\epsilon}{4}\big) \tag{5.100}$$

$$I_m(\frac{1}{2\epsilon}) \sim 1-\epsilon m^2+\frac{\epsilon}{4} \approx \exp\big(-\epsilon m^2+\frac{\epsilon}{4}\big)\ . \tag{5.101}$$

Using (5.100) and (5.101) in (5.78) we find that (5.81) is valid with the substitution

$$q = \mathrm{e}^{-\epsilon(\sigma+1)} \tag{5.102}$$

instead of (5.80). This version of Fay's solution becomes (Blackstock 1964)

$$V(\sigma,\theta) = 2\epsilon\sum_{n=1}^{\infty}\frac{\sin n\theta}{\sinh\big(n\epsilon(\sigma+1)\big)}\ , \tag{5.103}$$

which, like (5.85) also is an exact solution of (2.76). The only difference between the two versions is that σ in (5.85) has been replaced by $\sigma+1$ in (5.103). An improvement of Fay's solution (5.103), valid to the next order in ϵ, is given by Blackstock (1964) and Coulouvrat (1989).

Now we can compare the solution (5.103) with the Fourier expansion of the inviscid sawtooth solution. In the limit $\epsilon \to 0$ the Fourier coefficients in (5.103) approach the coefficients B_n given in (4.69). Obviously the higher harmonics in (5.103) are damped faster than the fundamental frequency wave with growing σ, so that, for $\sigma \gg 1$, (5.103) can be approximated as

$$V \approx 4\epsilon \exp(-\epsilon\sigma) \sin\theta \tag{5.104}$$

This result satisfies the heat conduction equation (3.65), obtained from Burgers' equation (2.76) by neglecting the nonlinear term. The knowledge of the exact solution gives us in this case the value of the amplitude constant in the asymptotic solution (5.104).

Inserting the definitions (5.2) of V, (5.6) of ϵ and (4.37) of σ, we find that the asymptotic fluid velocity does not depend on the initial velocity amplitude v_0. This is an example of the nonlinear wave phenomenon *saturation*. It means physically that a nonlinear medium has a limit of capacity of wave transmission.

The asymptotic behaviour of the Fay solution (5.103) for large σ including the higher harmonics is

$$V(\sigma, \theta) = 4\epsilon \sum_{n=1}^{\infty} \mathrm{e}^{-n\epsilon\sigma} \sin n\theta \ . \tag{5.105}$$

From (5.105) it is seen that the harmonics decay like $\mathrm{e}^{-n\epsilon\sigma}$. On the other hand, from (3.78) it is clear that the harmonics according to linear theory decays like $\mathrm{e}^{-n^2\epsilon\sigma}$. The difference is explained by the very different origin of higher harmonics ($n > 1$) considered in nonlinear and linear theory. In nonlinear theory the higher harmonics are generated by nonlinearity during the propagation of a wave, which begins as a pure sine wave. In linear theory the higher harmonics must be imposed by the boundary condition. Otherwise they are not there.

The somewhat curious conclusion may be drawn that the frequencies are decaying nonlinearly when the wave is at a very large distance, in the linear regime. This would mean that either the wave propagation is extremely sensitive to higher harmonics of very small amplitude, or that nonlinearity always must be taken into account - even when in linear regime. In either case the effect should be noticable so that linear theory also must be extremely sensitive to higher harmonics of very small amplitude, or that nonlinear theory always must be used. Fortunately, it is not necessary to assume either, and the explanation is simple. Prior to the regime where (5.104) holds, only the first harmonic is of importance. The total wave is a pure sine wave described equivalently by the linear result $e^{-n^2\epsilon\sigma}$, or the nonlinear result $e^{-n\epsilon\sigma}$, since $n^2 = n = 1$ for the first harmonic. (Contributions from higher harmonics, $n > 1$, are small enough to be neglected.)

5.3 Planar harmonic waves. The Khokhlov-Soluyan solution

5.3.1 Derivation of the Khokhlov-Soluyan solution

Fay's solution (5.103) is a refinement of the sawtooth wave (4.61) and (4.62), obtained by taking into account the nonzero value of ϵ. The discontinuity of (4.61) and (4.62) has been replaced by a shock with a finite width. However, the shock structure cannot be seen directly from (5.103). Another expression for the wave, exhibiting the shock structure, is now derived (Soluyan and Khokhlov 1961; Rudenko and Soluyan 1977, chapter 2) from the Cole-Hopf solution (3.71) using (3.81) for $U(\sigma,\theta)$. This expression will now be derived by the saddle-point method used in subsection 3.2.2 and compared to Fay's solution.

The boundary condition (5.64) together with (3.85) and (3.81) gives

$$U(\sigma,\theta) = \frac{C}{2\sqrt{\pi\epsilon\sigma}} \int_{-\infty}^{\infty} \exp(-\frac{\chi(\theta')}{2\epsilon})\, d\theta' \; , \qquad (5.106)$$

where, for the boundary condition $V_0 = \sin\theta$,

$$\chi(\theta') = \frac{(\theta-\theta')^2}{2\sigma} + \cos\theta' \; . \qquad (5.107)$$

We impose the condition

$$\chi'(\theta') = 0 \ , \tag{5.108}$$

which is satisfied for $\theta' = \theta'_m(\theta)$, where $\theta'_m(\theta)$ is given implicitly by

$$\theta - \theta'_m + \sigma \sin \theta'_m = 0 \ . \tag{5.109}$$

The first terms of the Taylor expansion of $\chi(\theta')$ about the extremum point $\theta' = \theta'_m$ is

$$\chi(\theta') = \frac{(\theta - \theta'_m)^2}{2\sigma} + \cos \theta'_m + \frac{1}{2}(\frac{1}{\sigma} - \cos \theta'_m)(\theta' - \theta'_m)^2 + \dots \ . \tag{5.110}$$

First we assume $\sigma < 1$. Then equation (5.109) has the solution $\theta'_m = \theta'_0$ and, as is seen from (5.107), χ has a minimum at $\theta' = \theta'_0$. This means that the integrand in (5.106) has a maximum. The integral (5.106) can now be evaluated, using (5.110):

$$U = C \frac{\exp[\left(-(\theta - \theta'_0)^2 - 2\sigma \cos \theta'_0 \right) / 4\epsilon\sigma]}{\sqrt{1 - \sigma \cos \theta'_0}} \ . \tag{5.111}$$

From (3.71) we obtain the approximate solution of Burgers' equation (2.76) using (5.111) and (5.109),

$$\begin{aligned} V &= -\frac{\theta - \theta'_0}{\sigma} + \frac{d\theta'_0}{d\theta}\left((\theta - \theta'_0)\frac{1}{\sigma} + \sin \theta'_0\right) + O(\epsilon) \\ &= -\frac{\theta - \theta'_0}{\sigma} + O(\epsilon) = \sin \theta'_0 + O(\epsilon) \ , \end{aligned} \tag{5.112}$$

where θ'_0 is the unique value of θ'_m which satisfies (5.109) for $\sigma < 1$. From (5.112) we find that

$$\theta = -\sigma V + \arcsin V + O(\epsilon) \ . \tag{5.113}$$

This is the same result as was derived in (4.36) for the lossless plane wave case.

However, when σ increases and passes the value $\sigma = 1$, the number of solutions of (5.109) increases to three, as is seen from Figure 5.2.

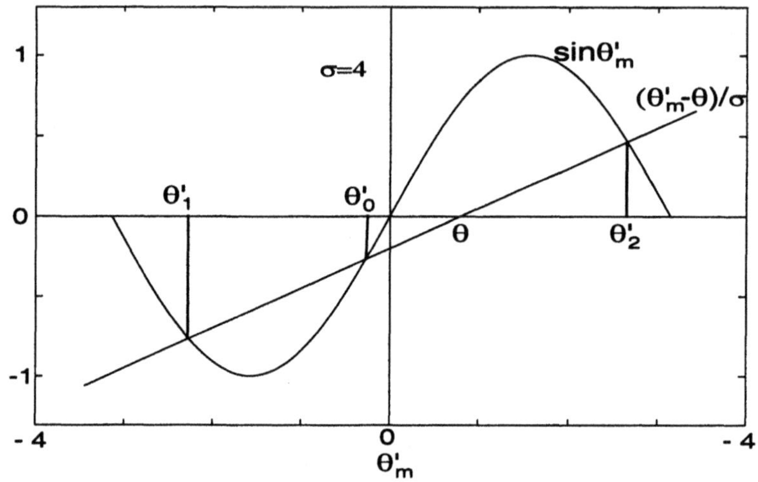

Figure 5.2: Three extremal points θ'_m exist when $\sigma > 1$.

From (5.110) it is seen that $\chi(\theta')$, has a minimum for θ' equal to the smallest and the largest root of (5.109). Thus the integration of (5.106) is taken over two maxima of the integrand and the result is obtained as a sum of two terms as in (5.111),

$$U = C\left(\frac{\exp(-\frac{\chi(\theta'_1)}{2\epsilon})}{\sqrt{1-\sigma\cos\theta'_1}} + \frac{\exp(-\frac{\chi(\theta'_2)}{2\epsilon})}{\sqrt{1-\sigma\cos\theta'_2}}\right) , \tag{5.114}$$

where θ'_1 and θ'_2 are the smallest and the largest root, respectively, of the equation (5.109). Using the result from (5.107) and (5.109)

$$\frac{\partial\chi(\theta'_{1,2})}{\partial\theta} = \frac{\theta-\theta'_{1,2}}{\sigma} = -\sin\theta'_{1,2} \ , \tag{5.115}$$

we obtain, using (3.71), the solution

$$V = 2\epsilon\frac{\partial U}{\partial\theta} = \frac{\exp(-\frac{\chi(\theta'_1)}{2\epsilon})\sin\theta'_1 + \exp(-\frac{\chi(\theta'_2)}{2\epsilon})\sin\theta'_2}{\exp(-\frac{\chi(\theta'_1)}{2\epsilon}) + \exp(-\frac{\chi(\theta'_2)}{2\epsilon})} \ . \tag{5.116}$$

In the derivation of (5.116) we have neglected the square roots in the denominators in (5.114). This can be done because the derivatives of the denominators in (5.114) are of $O(\epsilon)$ compared to the derivatives of the numerators and because of the fact that, for σ large enough,

$$\theta_2' - \theta_1' \approx 2\pi \ , \tag{5.117}$$

(see Figure 5.2). In order to simplify (5.116) we put

$$\begin{aligned} \theta_1' &= -\pi + \Phi_1 \\ \theta_2' &= \pi - \Phi_2 \ , \end{aligned} \tag{5.118}$$

where Φ_1 and Φ_2 are small. Because according to (5.109) we have

$$\theta = \theta_1' - \sigma \sin \theta_1' = \theta_2' - \sigma \sin \theta_2' \ , \tag{5.119}$$

writing

$$\sin \Phi_{1,2} \approx \Phi_{1,2} \tag{5.120}$$

we find from (5.118) and (5.119):

$$\begin{aligned} \Phi_1 &\approx (\pi + \theta)/(\sigma + 1) \\ \Phi_2 &\approx (\pi - \theta)/(\sigma + 1) \quad . \end{aligned} \tag{5.121}$$

Using (5.118) and inserting (5.121) into (5.116) we obtain

$$V = \frac{1}{\sigma + 1}\Big(- \theta + \pi \frac{\exp(\frac{\chi_1 - \chi_2}{4\epsilon}) - \exp(-\frac{\chi_1 - \chi_2}{4\epsilon})}{\exp(\frac{\chi_1 - \chi_2}{4\epsilon}) + \exp(-\frac{\chi_1 - \chi_2}{4\epsilon})}\Big) \ , \tag{5.122}$$

where we have put $\chi_{1,2} = \chi(\theta'_{1,2})$.

The parameters $\chi_1 - \chi_2$ remains to be found. Using (5.121) we obtain from (5.107) and (5.118) the result

$$\begin{aligned} \chi(\theta_1') - \chi(\theta_2') &= \cos \theta_1' - \cos \theta_2' + \frac{(\theta - \theta_1')^2}{2\sigma} - \frac{(\theta - \theta_2')^2}{2\sigma} \\ &\approx \frac{1}{2}(\Phi_1^2 - \Phi_2^2) + \frac{1}{2\sigma}\Big[(\pi - \Phi_1)^2 - (\pi - \Phi_2)^2 + 2\theta(2\pi - \Phi_1 - \Phi_2)\Big] \\ &= \frac{2\pi\theta}{\sigma + 1} \ . \end{aligned} \tag{5.123}$$

From (5.122) and (5.123) we obtain the Khokhlov-Soluyan solution (Soluyan and Khokhlov 1961)

$$V = \frac{1}{\sigma + 1}\Big(- \theta + \pi \tanh \frac{\pi\theta}{2\epsilon(\sigma + 1)}\Big) \ , \qquad -\pi < \theta < \pi \ . \tag{5.124}$$

5.3.2 Comparison between the Fay and the Khokhlov-Soluyan solutions

In contrast to Fay's solution (5.103), the Khokhlov-Soluyan solution (5.124) is not valid at arbitrarily large distance σ. This is seen from the fact that for increasing σ the straight line in Figure 5.2 becomes more horizontal and finally cuts the sine curve at more than three points. When this happens, the calculation of the solution (5.124), which presupposes only two minima of $\chi(\theta')$ is no more valid.

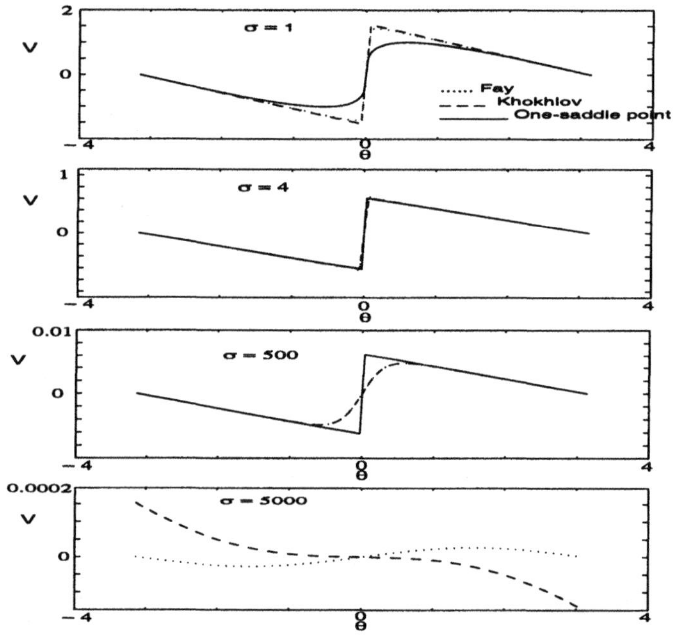

Figure 5.3. A comparison between the Fay solution (5.103), correct for $\sigma > 3$, the Khokhlov-Soluyan solution (5.124) and the one-saddle-point solution (4.79), correct for $(1+\sigma) \ll 1/\epsilon$. The dissipation $\epsilon = 10^{-7}$.

In Figure 5.3 is shown the Fay and Khokhlov-Soluyan solutions, and the non-dissipative solution exact solution. At $\sigma = 1$ the Fay and Khokhlov-Soluyan solutions are not expected to be correct. At $\sigma = 4$ and $\sigma = 500$ the Khokhlov-Soluyan and Fay are coinciding. The Khokhlov-Soluyan solution has a very weak dissipation dependence at first. Later, when the dissipation part is greater, the wave around

$\theta = 0$ dips first in the wrong way before being turned right by the non-dissipative term. It does not go through the points $-\pi$ and π whereas the Fourier series ansatz for the Fay solution makes the wave go through the points $\pm\pi$. So at $\sigma = 5000$ in Figure 5.3 the Khokhlov-Soluyan solution is not even close to periodic and gives a completely wrong behaviour. The number δ is defined as a measure of the shock width,

$$\delta = \frac{2\epsilon(\sigma+1)}{\pi} . \tag{5.125}$$

This number is assumed to be small since for θ passing from $-\delta$ to δ, the value of the function $\tanh\frac{\theta}{\delta}$ grows by the amount $2\tanh 1 \approx 1.523$. The condition for the validity of the Khokhlov-Soluyan solution (5.124) can be found by expanding it as a Fourier series

$$V = \sum_{k=1}^{\infty} d_k \sin k\theta , \tag{5.126}$$

where

$$\begin{aligned} d_k &= \frac{1}{\pi}\int_{-\pi}^{\pi} V \sin k\theta \, d\theta = \\ &= \frac{1}{\pi(\sigma+1)}\int_{-\pi}^{\pi}(-\theta + \pi\tanh\frac{\theta}{\delta})\sin k\theta\, d\theta . \end{aligned} \tag{5.127}$$

Following Hedberg (1994[a,b]), we expand

$$\begin{aligned} \tanh\frac{\theta}{\delta} &= 1 + 2\sum_{m=1}^{\infty}(-1)^m \mathrm{e}^{-2m\frac{\theta}{\delta}} \qquad & \frac{\theta}{\delta} > 0 \\ \tanh\frac{\theta}{\delta} &= -1 - 2\sum_{m=1}^{\infty}(-1)^m \mathrm{e}^{2m\frac{\theta}{\delta}} \qquad & \frac{\theta}{\delta} < 0 . \end{aligned} \tag{5.128}$$

By use of the integrals

$$\int_0^{\pi} \mathrm{e}^{-\alpha\theta}\sin k\theta\, d\theta = \left(1 - (-1)^k \mathrm{e}^{-\alpha\pi}\right)\frac{k}{\alpha^2 + k^2} \tag{5.129}$$

and

$$\int_0^{\pi} \theta \sin k\theta\, d\theta = (-1)^k \frac{2\pi}{k} , \tag{5.130}$$

where k is an integer, and the expansions (5.128), the Fourier coefficient d_k can be calculated:

$$
\begin{aligned}
d_k &= \frac{1}{\pi}\int_{-\pi}^{\pi} V \sin k\theta \, d\theta \\
&= \frac{1}{\pi(\sigma+1)}\int_{-\pi}^{\pi}(-\theta + \pi \tanh\frac{\theta}{\delta}) \sin k\theta \, d\theta \\
&= \frac{2}{(\sigma+1)k} + \frac{4}{\sigma+1}\sum_{m=1}^{\infty}\frac{k(-1)^m(1-(-1)^k e^{-2m\pi/\delta})}{(2m/\delta)^2+k^2} \,. \qquad (5.131)
\end{aligned}
$$

For

$$
e^{-2\pi/\delta} = \exp\left(\frac{-\pi^2}{\epsilon(\sigma+1)}\right) \ll 1 \,, \qquad (5.132)
$$

we neglect the second term in the parenthesis in the sum in (5.131) and obtain

$$
d_k \approx \frac{2}{\sigma+1}\left[\frac{1}{k} + 2k\sum_{m=1}^{\infty}\frac{(-1)^m}{(2m/\delta)^2+k^2}\right] . \qquad (5.133)
$$

The sum on the righthand side of (5.133) can be evaluated employing the representation of an analytic function in terms of its poles (Whittaker and Watson (1950) p. 134). To this end we consider a function $f(z)$ whose only singular points in the finite region of the complex plane are the poles at $a_1, a_2, a_3 \ldots$, where $|a_1| \le |a_2| \le |a_3| \le \ldots$. Let the residue of these poles be $b_1, b_2, b_3 \ldots$ and assume that it is possible to choose a sequence of circles C_m with radii R_m and centres at $x = 0$ which does not cross anyone of the poles. This means that $|f(z)|$ is bounded on C_m. Assume that, for all points on C_m, $|f(z)| < M$ where M is independent of m. Then we have, according to the residue theorem,

$$
\frac{1}{2\pi i}\int_{C_m}\frac{f(z)}{z-x}dz = f(x) + \sum_r \frac{b_r}{a_r - x} \,, \qquad (5.134)
$$

where the sum is taken over all poles within C_m. If $f(z)$ is analytic at $z = 0$ we obtain from (5.134) writing $\frac{1}{z-x}$ as $\frac{1}{z} + \frac{x}{z(z-x)}$:

$$
\begin{aligned}
\frac{1}{2\pi i}\int_{C_m}\frac{f(z)}{z-x}dz &= \frac{1}{2\pi i}\int_{C_m}\frac{f(z)}{z}dz - \frac{x}{2\pi i}\int_{C_m}\frac{f(z)}{z(z-x)}dz \\
&= f(0) + \sum_r \frac{b_r}{a_r} + \frac{x}{2\pi i}\int_{C_m}\frac{f(z)}{z(z-x)}dz \,, \qquad (5.135)
\end{aligned}
$$

where the sum is taken over all the poles within C_m. For $m \to \infty$ we have

$$\Big| \int_{C_m} \frac{f(z)}{z(z-x)} dz \Big| \leq \frac{2\pi M}{R_m} . \tag{5.136}$$

Because $\lim_{m\to\infty} R_m^{-1} = 0$ we obtain by letting $m \to \infty$ in (5.135), and using (5.134) the result

$$f(x) = f(0) + \sum_{n=1}^{\infty} b_n \Big(\frac{1}{x - a_n} + \frac{1}{a_n} \Big) . \tag{5.137}$$

We shall now apply (5.137) to the function

$$f(z) = \frac{1}{\sin z} - \frac{1}{z} . \tag{5.138}$$

The singularities of $f(z)$ are situated at the points $z = n\pi$, $n = \pm 1, \pm 2, \pm 3, \ldots$. The residues of $f(z)$ at the pole $z = n\pi$ is $(-1)^n$ and it is obvious that $f(z)$ is bounded on the circles $|z| = (n + \frac{1}{2})\pi$ as $n \to \infty$. Thus in view of the general result (5.137) and

$$f(0) = \lim_{z \to 0} \frac{z - \sin z}{z \sin z} = 0 , \tag{5.139}$$

we obtain

$$\frac{1}{\sin z} - \frac{1}{z} = \sum_{n=\pm 1, \pm 2, \ldots} (-1)^n \Big(\frac{1}{z - n\pi} + \frac{1}{n\pi} \Big) . \tag{5.140}$$

Using the fact that the terms $1/n\pi$ in (5.140) cancel in pairs and that $\sinh z = -i \sin iz$ we obtain from (5.140),

$$\frac{1}{\sinh z} = \frac{1}{z} + 2z \sum_{n=1}^{\infty} \frac{(-1)^n}{z^2 + (n\pi)^2} . \tag{5.141}$$

When applied to the sum in (5.133) the result (5.141) yields

$$d_k = \frac{2\epsilon}{\sinh \big(k\epsilon(\sigma + 1) \big)} , \tag{5.142}$$

in agreement with (5.103).

Since Fay's solution is valid for arbitrarily large distance σ, the condition (5.132) which is the only assumption in the approximate Fourier expansion of the Khokhlov-Soluyan solution, restricts its validity in distance.

Novikov (1978) has given an expression of a more general wave form than Khokhlov-Soluyan, based on sums of discrete heat source solutions of the heat conduction solution.

Using ϑ-functions, Fay's and Khokhlov-Soluyan's solutions are compared with respect to the corresponding solutions to the heat conduction equation by D.F. Parker (1980).

An expression of the periodic solution of Burgers' equation in the from of an amplitude modulated sine wave is given by A. Parker (1992).

Asymptotic methods connecting solutions in the shock formation region with Fay's solution are given by Coulouvrat (1991[a]).

5.3.3 Comparison between the Khokhlov-Soluyan solution and the sawtooth solution

It is also interesting to investigate the relation between the Khokhlov-Soluyan solution and the fully developed sawtooth wave given by (4.61) and (4.62) as solution of the Riemann wave equation (4.48). The following considerations are analogous to those for the matching of the N-wave solutions (5.22) and (5.23).

First we state (cf. (4.48) and (4.60)) that the solution function V_0 given implicitly by

$$V_0 = \sin(\theta + \sigma V_0) \tag{5.143}$$

is the first term in the outer series expansion

$$V = V_0 + \epsilon V_1 + \dots \tag{5.144}$$

of the solution of (2.76). From (5.143) we find the results

$$\theta = \arcsin V_0 - \sigma V_0 \tag{5.145}$$

and

$$\frac{d\theta}{dV_0} = \frac{1}{\sqrt{1 - V_0^2}} - \sigma \; . \tag{5.146}$$

For $V_0 = 0$, $\sigma = 1$, we obtain

$$\left(\frac{\partial V_0}{\partial \theta}\right)_{\theta=0} = \infty \, . \tag{5.147}$$

The infinite derivative of $V_0(\sigma, \theta)$ at $(\sigma = 1, \theta = 0)$ is the beginning of the discontinuity of the wave profile. In order to find a solution in the neighbourhood of $\theta = 0$ we "stretch" the θ variable and define

$$\theta^* = \frac{\theta}{\epsilon} \, . \tag{5.148}$$

This stretching is natural, since V changes sharply in a short θ-interval in the neighbourhood of $\theta = 0$. Insertion of (5.148) into (2.76) gives the following equation for $V^*(\sigma, \theta^*) = V(\sigma, \theta)$:

$$\epsilon \frac{\partial V^*}{\partial \sigma} - V^* \frac{\partial V^*}{\partial \theta^*} - \frac{\partial^2 V^*}{\partial \theta^{*2}} = 0 \, , \tag{5.149}$$

Substituting the "inner" expansion

$$V^* = V_0^* + \epsilon V_1^* + \ldots \tag{5.150}$$

into (5.149) and integrating the equation for V_0^* we have

$$\frac{V_0^{*2}}{2} + \frac{\partial V_0^*}{\partial \theta^*} = \frac{1}{2} C^2(\sigma) \, , \tag{5.151}$$

where $C(\sigma)$ is an arbitrary function. Integration of (5.151) gives

$$\theta^* = \frac{1}{C} \ln \left| \frac{V_0^* + C}{V_0^* - C} \right| + D \, , \tag{5.152}$$

where D depends on σ. The derivative $\frac{\partial V_0}{\partial \theta}$ in (5.146) at the discontinuity at $\theta = 0$ becomes infinite through increasing positive values. Therefore we assume $\frac{\partial V_0^*}{\partial \theta^*} > 0$, that is, $C > V_0^*$. Thus we can solve V_0^* in (5.152):

$$V_0^* = C \tanh\left(\frac{C}{2}(\theta^* - D)\right) . \tag{5.153}$$

We immediately find that $D = 0$ since the solution, like the initial wave (5.64), must be antisymmetric with respect to θ. The constant C, on the other hand, is determined by the asymptotic matching in the same way as the undetermined constant in (5.12). In analogy with (5.14) we have

$$\lim_{\theta \to 0_+} V_0^* = \lim_{\theta^* \to \infty} V_0 \ , \tag{5.154}$$

which on using (5.143) and (5.153) gives

$$C = \sin p_0, \qquad p_0 = \sigma V_0 \ , \tag{5.155}$$

where p_0 is given by the transcendental equation,

$$p_0 = \sigma \sin p_0 \ . \tag{5.156}$$

For $\sigma \gg 1$ the value of p_0 approaches π. So we rewrite (5.156) with $p_0 = \pi - \delta$:

$$\pi - \delta = \delta \sin(\pi - \delta) = \sigma \sin \delta \approx \sigma\delta \ . \tag{5.157}$$

Thus we obtain

$$\delta = \frac{\pi}{1 + \sigma} \tag{5.158}$$

and

$$C = \sin p_0 = \sin \delta \approx \delta = \frac{\pi}{1+\sigma} \ . \tag{5.159}$$

The solution of (5.143) can be approximated by a straight line for $\sigma \gg 1$. This straight line is given by (4.62) for $-\pi < \theta < 0$ and by (4.61) for $0 < \theta < \pi$. This is the outer solution. The inner solution (5.153) with C given by (5.159) can be combined with the outer solution to give

$$V = \frac{\pi}{\sigma + 1}\Big(\tanh\big(\frac{\pi\theta}{2\epsilon(\sigma+1)}\big) - \frac{\theta}{\pi}\Big) \ . \tag{5.160}$$

This result is obtained by adding the solutions mentioned and subtracting their common part in the limit $\epsilon \to 0$. In (5.160) we recognize the Khokhlov-Soluyan solution (5.124), obtained earlier by a different route.

5.4 Planar harmonic waves. The exact solution

5.4.1 Recursion formulae for the Fourier series of the exact solution

The problem of finding the evolution of a planar harmonic sound wave with nonlinear effects in a weakly dissipative medium has been extensively studied in the present book. Putting the dissipation equal to zero we have found the Bessel-Fubini solution (4.46) and the sawtooth solution (4.61), (4.62). Formula (4.67) shows how the Fourier coefficients of these two solutions transform into each other; for $\sigma < 1$ the first term of (4.67) is zero, and for $\sigma \to \infty$ the second term of (4.67) vanishes. However, for nonzero dissipation, Fay's solution (5.85) which replaces the sawtooth solution (4.61), (4.62), is useful only for $\sigma \gg 1$, as was explained earlier (see (5.81)). The Bessel-Fubini solution, on the other hand, is valid in the region where the wave is continuous, i.e. for $\sigma < 1$. For $\epsilon \neq 0$ the Bessel-Fubini solution can be considered as the first term of an asymptotic expansion of the solution in powers of ϵ. Fay's solution can be considered in the same way. Thus by means of perturbation methods, corrections of the order ϵ can be found for both the Fay solution and for the Bessel-Fubini solution (Blackstock 1964). The corrected Bessel-Fubini solution is

$$V(\sigma,\theta) = \sum_{m=1}^{\infty}[\frac{2J_m(m\sigma)}{m\sigma} - \frac{2\epsilon}{m}\sum_{r=1}^{\infty} ra_r(J_{m-r}(m\sigma) + J_{m+r}(m\sigma))]\sin m\theta \tag{5.161}$$

where

$$a_r = r(1-\sigma^2)^{-\frac{1}{2}}\sigma^{-r}[1-(1-\sigma^2)^{\frac{1}{2}}]^r \tag{5.162}$$

However, the problem of finding a representation of the solution valid for arbitrary σ and $\epsilon \neq 0$ remains. In this section such a representation will be found using the exact system of (5.78). From this, Fay's solution (5.85) with the Fourier coefficients (5.92) can be found by (approximately) putting the modified Bessel functions equal in (5.78).

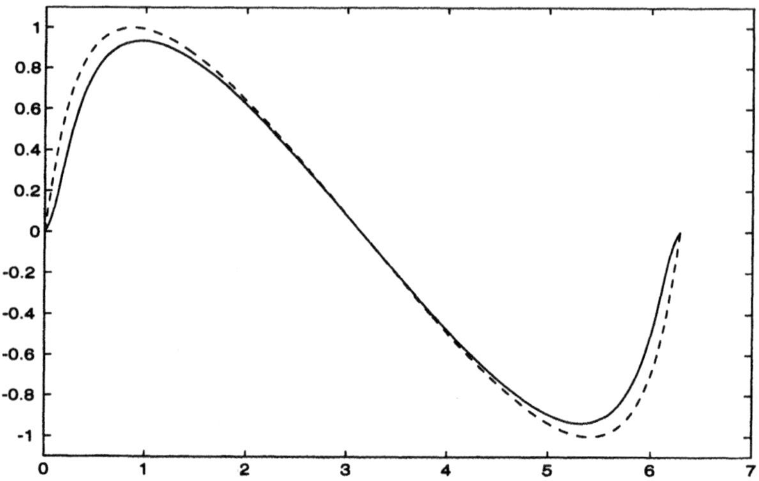

Figure 5.4: A comparison between the Bessel-Fubini solution (dashed) and the corrected Bessel-Fubini solution (solid) at $\sigma = 0.7$ and with $\epsilon = 0.1$.

The Fourier coefficients a_m in (5.92) are expanded in the series

$$a_m = 4\epsilon e^{-m\epsilon\sigma}(1 + e^{-2m\epsilon\sigma} + e^{-4m\epsilon\sigma} + \ldots) \ . \tag{5.163}$$

An obvious generalization of (5.163) is the expansion of the exact Fourier coefficients a_m in (5.78) (Enflo and Hedberg 2001)

$$a_k = 4\epsilon \sum_{r=1,2\ldots}^{\infty} d_{kr} e^{-r\epsilon\sigma} \ . \tag{5.164}$$

The limit corresponding to Fay's solution would require

$$d_{kr} \to 1 \qquad r = (2n+1)k\,, \quad n = 0, 1, 2\ldots \tag{5.165}$$

$$d_{kr} \to 0 \qquad r \neq (2n+1)k\,, \quad n = 0, 1, 2\ldots \tag{5.166}$$

If (5.164) is inserted into (5.78) we obtain, using (5.80),

$$I_m(\frac{1}{2\epsilon})m$$
$$= \sum_{k=1}^{\infty}(-1)^{k+1} q^{k^2} \Big\{ q^{-2km} I_{k-m}(\frac{1}{2\epsilon})(d_{k1}q + d_{k2}q^2 + \ldots + d_{k,4km}q^{4km})$$

$$+ q^{2km}\Big([I_{k-m}(\frac{1}{2\epsilon})d_{k,4km+1} - I_{k+m}(\frac{1}{2\epsilon})d_{k1}]q$$
$$+ [I_{k-m}(\frac{1}{2\epsilon})d_{k,4km+2} - I_{k+m}(\frac{1}{2\epsilon})d_{k2}]q^2 + \dots\Big)\Big\}$$
$$m = 1, 2, \dots \ . \tag{5.167}$$

Every positive integer value of m in (5.167) gives for each power of q, a linear equation for the coefficients in the series (5.164).

By a simple argument we shall show that some of the coefficients d_{ik} are zero. According to (5.167) the exponent r in q^r is given as

$$r = k^2 \pm 2km + s = (k \pm m)^2 + s - m^2 \ . \tag{5.168}$$

For given r and m given, the possible solutions of (5.168) yield the numbers k, s in the coefficients d_{ks} which occur in the equation obtained by putting the coefficients of q^r equal on both sides of (5.167). From (5.168) it is obvious that for r even, $k+s$ must be even and for r odd, $k+s$ must be odd. Thus, d_{ks} with even and odd value of $k+s$ respectively, occur in different equations. Inhomogeneous terms in the system of equations (5.167) for d_{ks} occur only for $r = 0$. i.e. even. Thus we obtain a homogeneous system for d_{ks} with $k+s$ odd. This system has the only solution $d_{ks} = 0$. Henceforth we assume $k+s$ to be even.

Another conclusion from (5.167) is that d_{ks} is zero for $k > s$. To see this we first note from (5.167) that

$$d_{11} = \frac{I_1}{I_0} \ . \tag{5.169}$$

On the other hand, from (5.168) the lowest value of the exponential r of q^r in (5.167) is (put $k = m$, $s = 1$)

$$r_{min} = 1 - m^2 \ . \tag{5.170}$$

Using (5.170) we conclude from (5.167):

$$d_{m1} = 0 \ , \qquad m > 1 \ . \tag{5.171}$$

The next lowest value of r is $2 - m^2$. From the terms in (5.167) contributing to this power of q we conclude, using the relation $I_{-m} = I_m$:

$$I_0 d_{m2} - I_1(d_{m+1,1} + d_{m-1,1}) = 0 \ . \tag{5.172}$$

If we assume $m > 2$ in (5.172) and use (5.171) we find:

$$d_{m2} = 0 \ , \qquad m > 2 \ . \tag{5.173}$$

The complete recursion formula for the coefficients d_{ms}, a generalization of (5.172), is (Enflo and Hedberg 2001)

$$\begin{aligned} I_0 d_{ms} &= I_1(d_{m-1,s-1} + d_{m+1,s-1}) - I_2(d_{m-2,s-4} + d_{m+2,s-4}) \\ &+ \ \ldots + (-1)^{n+1} I_n(d_{m-n,s-n^2} + d_{m+n,s-n^2}) + \ldots \ , \end{aligned} \tag{5.174}$$

where the following holds:

(1) $d_{0k} = d_{k0} = 0$ with the exception $d_{00} = 1$.
(2) $d_{-m,s} = -d_{ms}$.
(3) The series on the righthand side of (5.174) terminates with $(-1)^{r+1} r I_r$ if $s = r^2$.
Otherwise the series is broken for $n^2 > s$.
(4) $d_{ms} = 0$ for $m + s =$ odd integer.
(5) $d_{ms} = 0$ for $m > s$.

We have assumed $m, r \geq 1$ in (5.167), where r is the exponent in q^r.

Using the validity of (5.171), the rule (5) follows immediately from (5.174) by induction. All the d-functions on the righthand side of (5.174) can be calculated before d_{ms} is calculated. Because $m \leq s$, we can use (5.174) for calculating $d_{(m+s)/2,(m+s)/2}, d_{(m+s)/2-1,(m+s)/2+1}, \ldots$ $\ldots, d_{m-1,s+1}$ as also for calculating d_{lr} with $l+r < m+s$. Thus (5.174) gives the complete solution of the problem of finding the coefficients d_{kr} in the expansion (5.164); hence the problem of finding the Fourier coefficients in the expansion (5.73) of $V(\sigma, \theta)$ for all values of σ. It is possible to show that the solution of (5.174) when putting all I-functions equal and assuming $d_{00} = 1$, is given by (5.165) and (5.166).

5.4.2 Solving recursion formulae by discrete integration

Analytical expressions for the coefficients d_{ms} can be obtained from (5.174) by a procedure called discrete integration, which will now be illustrated. First we observe that (5.174), together with the rules for its application, directly gives

$$I_0 d_{rr} = I_1 d_{r-1,r-1} , \tag{5.175}$$

which yields the general result for d_{rr}:

$$d_{rr} = \frac{I_1^r}{I_0^r} . \tag{5.176}$$

The recursion formula for $d_{r,r+2}$ is found from (5.174):

$$I_0 d_{r,r+2} = I_1(d_{r-1,r+1} + d_{r+1,r+1}) - I_2 d_{r-2,r-2} . \tag{5.177}$$

Using (5.176) we find from (5.177) that

$$d_{r,r+2} - \frac{I_1}{I_0} d_{r-1,r+1} = \frac{I_1^{r+2}}{I_0^{r+2}} - \frac{I_2 I_1^{r-2}}{I_0^{r-1}} . \tag{5.178}$$

The concept of discrete integrals can be used if for a function $F(n)$ the difference

$$F(n) - F(n-1) = g(n) \tag{5.179}$$

is known. It is easy to see that the discrete integral of

$$f_k(n) \equiv n(n+1)\dots(n+k-1) = (n)_k \tag{5.180}$$

is

$$F_k(n) \equiv \frac{1}{k+1} n(n+1)\dots(n+k) = \frac{1}{k+1}(n)_{k+1} , \tag{5.181}$$

since

$$\begin{aligned} &F_k(n) - F_k(n-1) \\ &\quad = \frac{1}{k+1} n(n+1)\dots(n+k) - \frac{1}{k+1}(n-1)n\dots(n+k-1) \\ &\quad = \frac{1}{k+1} n(n+1)\dots(n+k-1)(n+k-n+1) = (n)_k . \end{aligned} \tag{5.182}$$

The functions $d_{r,r+2}$ and $\frac{I_1}{I_0}d_{r-1,r+1}$ differ only in their dependence on r. Thus, using (5.181) with $k = 0$ and $(n)_0 = 1$ we carry out a discrete integration of (5.178) and find that

$$d_{r,r+2} = r\left(\frac{I_1^{r+2}}{I_0^{r+2}} - \frac{I_2 I_1^{r-2}}{I_0^{r-1}}\right) \tag{5.183}$$

Only for $r = 1$, the term with I_2 is absent in (5.177) and we obtain

$$d_{13} = \frac{I_1^3}{I_0^3} \tag{5.184}$$

In order to write expressions for the other coefficients d_{ms} we introduce new notation and denote discrete integration by the operator S:

$$\begin{aligned}
&R_0 = 1\\
&R_1(r) = SR_0 = r\\
&R_2(r;k) = SR_1(r+k) = S(r+k) = \frac{1}{2}(r)_2 + k(r)_1\\
&R_3(r;k,l) = SR_2(r+l;k)\\
&R_4(r;k,l,m) = SR_3(r+m;k,l)\\
&R_{n+1}(r;k_1,\ldots,k_n) = SR_n(r+k_n;k_1,\ldots,k_{n-1})\,.
\end{aligned} \tag{5.185}$$

All discrete integrations denoted by the operator S are carried out in the variable r with the "integration constant" put equal to zero.

Using the notation (5.185) and the additional notation

$$\begin{aligned}
&\hat{R}_3(r;k,l) = R_3(r;k,l) + R_3(r;l,k)\\
&\hat{R}_4(r;k,k,l) = R_4(r;k,k,l) + R_4(r;k,l,k) + R_4(r;l,k,k)
\end{aligned} \tag{5.186}$$

and so on, we can now, using (5.174) and making successive discrete integrations starting from (5.178), write some more d_{ms} coefficients in a compact form:

$$d_{r,r+4} = \frac{I_1^{r+4}}{I_0^{r+4}}R_2(r;1) - \frac{I_2 I_1^{r}}{I_0^{r+1}}[R_2(r;1) + R_2(r;-2)] + \frac{I_2^2 I_1^{r-4}}{I_0^{r-2}}R_2(r;-2) \tag{5.187}$$

$$
\begin{aligned}
d_{r,r+6} &\\
= \frac{I_1^{r+6}}{I_0^{r+6}} R_3(r;1,1) &- \frac{I_2 I_1^{r+2}}{I_0^{r+3}} [R_3(r;1,1) + \hat{R}_3(r;1,-2) + R_1] \\
&+ \frac{I_2^2 I_1^{r-2}}{I_0^{r}} [\hat{R}_3(r;1,-2) + R_3(r;-2,-2)] - \frac{I_2^3 I_1^{r-6}}{I_0^{r-3}} R_3(r;-2,-2) \\
&+ \frac{I_3 I_1^{r-3}}{I_0^{r-2}} R_1
\end{aligned}
\tag{5.188}
$$

$$
\begin{aligned}
d_{r,r+8} &\\
= \frac{I_1^{r+8}}{I_0^{r+8}} &R_3(r;1,1,1) \\
&- \frac{I_2 I_1^{r+4}}{I_0^{r+5}} [R_4(r;1,1,1) + \hat{R}_4(r;1,1,-2) + R_2(r;1) + R_2(r;2)] \\
&+ \frac{I_2^2 I_1^{r}}{I_0^{r+2}} [\hat{R}_4(r;1,1,-2) + \hat{R}_4(r;1,-2,-2) + R_2(r;-2) + R_2(r;2)] \\
&- \frac{I_2^3 I_1^{r-4}}{I_0^{r-1}} [\hat{R}_4(r;1,-2,-2) + R_4(r;-2,-2,-2)] \\
&+ \frac{I_2^4 I_1^{r-8}}{I_0^{r-4}} \hat{R}_4(r;-2,-2,-2) + \frac{I_3 I_1^{r-1}}{I_0^{r}} [R_2(r;1) + R_2(r;-3)] \\
&- \frac{I_3 I_2 I_1^{r-5}}{I_0^{r-3}} [R_2(r;-2) + R_2(r;-3)]
\end{aligned}
\tag{5.189}
$$

$$
\begin{aligned}
&d_{r,r+10} \\
&= \frac{I_1^{r+10}}{I_0^{r+10}} R_5(r;1,1,1,1) \\
&- \frac{I_2 I_1^{r+6}}{I_0^{r+7}} [R_5(r;1,1,1,1) + \hat{R}_5(r;1,1,1,-2) + R_3(r;1,1) + \hat{R}_3(r;1,2)] \\
&+ \frac{I_2^2 I_1^{r+2}}{I_0^{r+4}} [\hat{R}_5(r;1,1,1,-2) + \hat{R}_5(r;1,1,-2,-2) \\
&\qquad + \hat{R}_3(r;1,-2) + \hat{R}_3(r;1,2) + \hat{R}_3(r;2,-2)]
\end{aligned}
$$

$$
\begin{aligned}
&- \frac{I_2^3 I_1^{r-2}}{I_0^{r+1}}[\hat{R}_5(r;1,1,-2,-2) + \hat{R}_5(r;1,-2,-2,-2) \\
&\qquad + \hat{R}_3(r;2,-2) + R_3(r;-2,-2)] \\
&+ \frac{I_2^4 I_1^{r-6}}{I_0^{r-2}}[\hat{R}_5(r;1,-2,-2,-2) + R_5(r;-2,-2,-2,-2)] \\
&- \frac{I_2^5 I_1^{r-10}}{I_0^{r-5}} R_5(r;-2,-2,-2,-2) \\
&+ \frac{I_3 I_1^{r+1}}{I_0^{r+2}}[R_3(r;1,1) + \hat{R}_5(r;1,-3)] \\
&- \frac{I_3 I_2 I_1^{r-3}}{I_0^{r-1}}[\hat{R}_3(r;1,-2) + \hat{R}_3(r;1,-3) + \hat{R}_3(r;-2,-3)] \\
&+ \frac{I_3 I_2^2 I_1^{r-7}}{I_0^{r-4}}[R_3(r;-2,-2) + \hat{R}_3(r;-2,-3)]
\end{aligned}
\tag{5.190}
$$

A negative power of an I-function in (5.187)-(5.190) means that the corresponding term is absent.

As an example we find from (5.185), (5.186) and (5.189)

$$
d_{2,10} = 42\frac{I_1^{10}}{I_0^{10}} - 52\frac{I_2 I_1^6}{I_0^7} + 9\frac{I_2^2 I_1^2}{I_0^4} + 2\frac{I_3 I_1}{I_0^2} \; . \tag{5.191}
$$

Every possible way of writing the number 10 as a sum of squares constitutes a term in $d_{2,10}$: $10 = 1^2 + \ldots + 1^2 = 2^2 + 1^2 + \ldots 1^2 = 2^2 + 2^2 + 1^2 + 1^2 = 3^2 + 1^2$. The limit of (5.191) giving Fay's solution is obtained from the asymptotic expression (5.79) of $I_n(\frac{1}{2\epsilon})$. For $\epsilon n^2 \ll 1$ we have

$$
\frac{I_n(\frac{1}{2\epsilon})}{I_0(\frac{1}{2\epsilon})} = 1 + O(\epsilon) \; . \tag{5.192}
$$

In our example we get

$$
d_{2,10} = 1 + O(\epsilon) \; , \tag{5.193}
$$

which is in agreement with the rule (5.165).

5.4.3 Comparison of Fourier coefficients in the Bessel-Fubini solution, the Fay solution and the exact solution

As a comparison between the Bessel-Fubini solution, the Fay solution and the exact solution, the coefficients $b_1, \ldots, b_4$ have been plotted in Figures 5.5 and 5.6. The recursion formula (5.166) has been used for the calculation of the Fourier coefficients $b_m(\epsilon, \sigma)$ in (5.73) according to the expansion (5.164) for $\sigma > 0.95$ and $\epsilon = 0.05$, 0.1. For the non-dissipative curves we have used the one saddle-point solution (4.79), an equivalent to Blackstock's (1966) expression (4.67) for $\epsilon = 0$. The figures 5.5 and 5.6 show a refinement of Blackstock's (1966) Figure 4, where he shows the corresponding Fourier coefficients in the case of zero dissipation. The calculation of the coefficients b_1 to b_4 has good numerical convergence for $\sigma > 0.9$, which means smooth transition to the Bessel-Fubini coefficients corrrected for dissipation (5.161). Curves similar to Blackstock's (1966) were also given by Cook (1962).

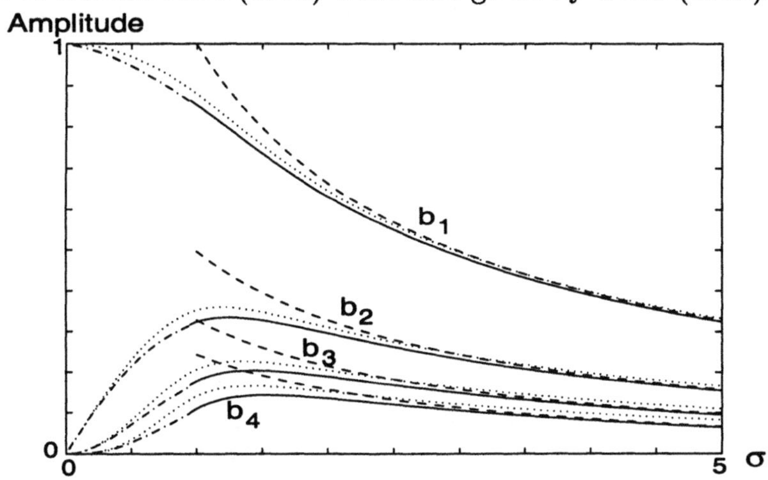

Figure 5.5: The coefficients $b_1 - b_4$ according to: the exact solution (5.73) with (5.164) are shown as solid lines; the Fay solution (5.85) are shown as dashed lines; and the corrected Bessel-Fubini (5.161) are shown as dash-dotted lines, all for $\epsilon = 0.05$. The non-dissipative solution ($\epsilon = 0$) - dotted.

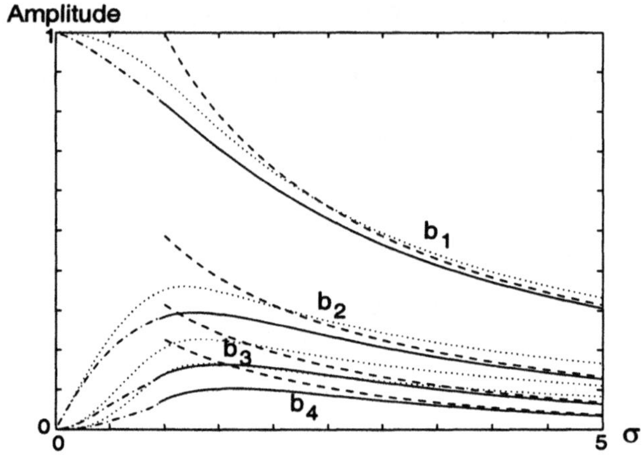

Figure 5.6: The coefficients $b_1 - b_4$ according to: the exact solution (5.73) with (5.164) are shown as solid lines; the Fay solution (5.85) are shown as dashed lines; and the corrected Bessel-Fubini (5.161) are shown as dash-dotted lines, all for $\epsilon = 0.1$. The non-dissipative solution ($\epsilon = 0$) - dotted.

An interesting consequence of the present analysis is the possibility of finding the maximum of each Fourier coefficient $b_2, b_3 \ldots$ as a function of σ. From Figure 5.7 and Table 5.1 it is seen that the maxima rise with both increasing dissipation and growing harmonics.

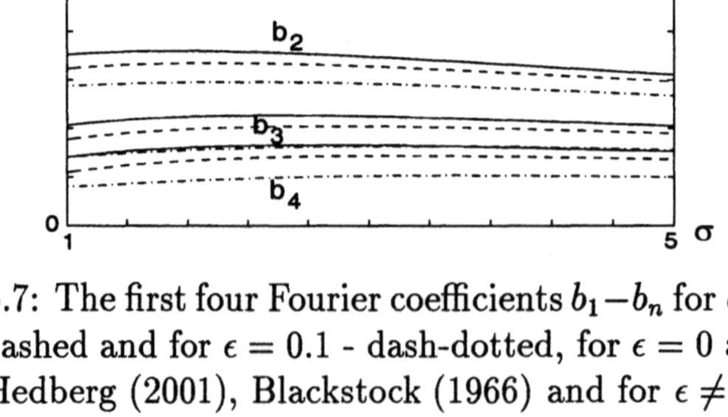

Figure 5.7: The first four Fourier coefficients $b_1 - b_n$ for $\epsilon = 0$ - solid, $\epsilon = 0.05$ - dashed and for $\epsilon = 0.1$ - dash-dotted, for $\epsilon = 0$ according to Enflo and Hedberg (2001), Blackstock (1966) and for $\epsilon \neq 0$ according to (5.164).

Maxima for $\epsilon = 0$:	
$b_2 = 0.3609$	at $\sigma = 1.18$
$b_3 = 0.2268$	at $\sigma = 1.34$
$b_4 = 0.1660$	at $\sigma = 1.41$
Maxima for $\epsilon = 0.05$:	
$b_2 = 0.3355$	at $\sigma = 1.24$
$b_3 = 0.2043$	at $\sigma = 1.43$
$b_4 = 0.1438$	at $\sigma = 1.53$
Maxima for $\epsilon = 0.1$:	
$b_2 = 0.2957$	at $\sigma = 1.30$
$b_3 = 0.1639$	at $\sigma = 1.53$
$b_4 = 0.1032$	at $\sigma = 1.63$

Table 5.1: The position of the maxima of higher harmonics increases with the number of harmonics and with dissipation ϵ.

Figures 5.8 and 5.9 give for one period the exact Mendousse solution (5.71), the Bessel-Fubini (4.46) and the present solution for $\sigma = 0.95$ and $\sigma = 5$ respectively. A small numerical irregularity is observed for the case $\sigma = 0.95$.

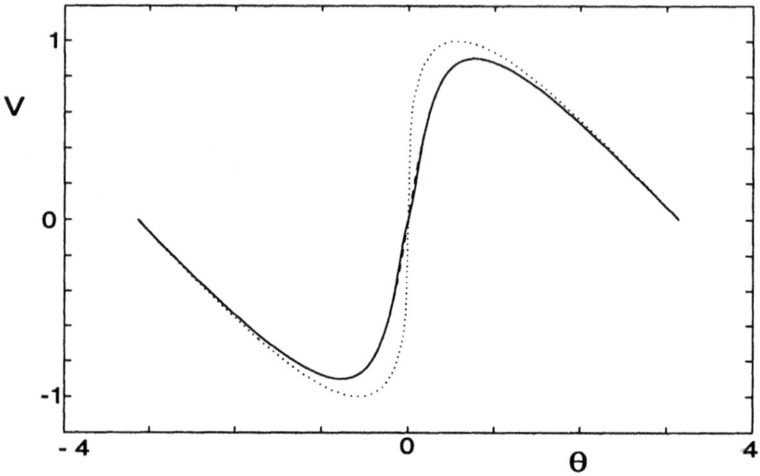

Figure 5.8: Solution over one period: dashed - the exact solution (5.71); dotted - the non-dissipative solution; solid - the present solution; at $\sigma = 0.95$.

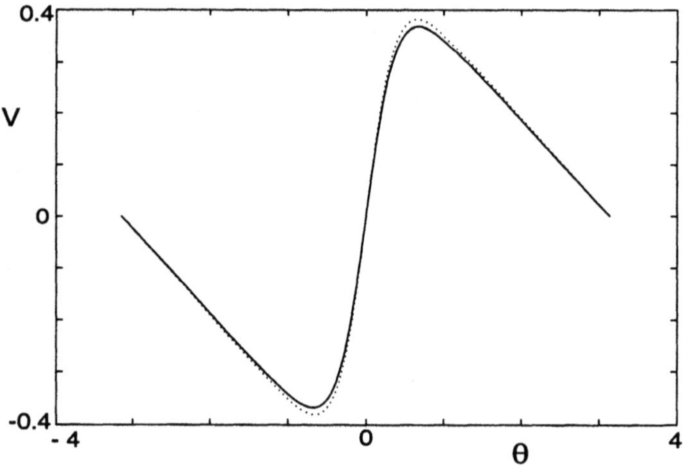

Figure 5.9: Solution over one period: dashed - the exact solution (5.71); dotted - the Fay solution (5.85); solid - the present solution; at $\sigma = 5$.

5.5 Multifrequency waves

5.5.1 Expressions for multifrequency solutions

A solution for multifrequency plane waves propagating through a dissipative and nonlinear medium will be presented, where any frequencies, amplitudes and phases may be chosen as input. The result is a generalization of the Mendousse (1953) solution (5.71) for one single frequency and of Lardner's (1982) solution for two frequencies. The method is based upon the Mendousse solution for a pure sinusiodal wave expressed as a ratio of Bessel function series found earlier in equation (5.71). The solution for multiple frequencies is calculated from the knowledge that the solution has a specified form, and that the unknown coefficients in this form can be extracted at zero distance from the sum of the single frequency solutions. As the single frequency solutions are exact, so is the solution for the multiple frequency boundary condition. The solution depends on the separation of variables, which means that once the coefficients are calculated, a solution valid at any distance is at once known.

Let us start with the statement that any solution for a periodic boundary condition may have form

$$U_{mfq} = \sum_{k=-\infty}^{\infty} c_k \mathrm{e}^{-k^2 \epsilon\sigma} \mathrm{e}^{ik\theta} , \tag{5.194}$$

(cf. (3.78)), where the subscript mfq stands for *multifrequency.* This is equivalent to the solution

$$V_{mfq}(\sigma,\theta) = 2i\epsilon \frac{\sum_{k=-\infty}^{\infty} k\, c_k\, \mathrm{e}^{-k^2\epsilon\sigma}\, \mathrm{e}^{ik\theta}}{\sum_{k=-\infty}^{\infty} c_k \mathrm{e}^{-k^2\epsilon\sigma} \mathrm{e}^{ik\theta}} \tag{5.195}$$

Let (5.195) be the solution to (2.76) with the multifrequency boundary condition involving L frequencies:

$$V_{mfq}(\sigma = 0, \theta) = \sum_{l=1}^{L} a_l \sin(b_l \omega_0 \theta + \phi_l) , \tag{5.196}$$

where b_ls are integers. Each one of the L frequencies has on their own a known single frequency solution V_l in a form similar to (5.71).

Because this is a nonlinear problem the single frequency solutions cannot be superposed to form the solution to the multiple frequency boundary condition – except at the distance $\sigma = 0$. At this distance no nonlinear (or linear) evolution has taken place, and superposition is allowed. So this is the σ-value at which the coefficients of the multifrequency wave will be extracted from the known coefficients of the single frequencies.

First some care must be taken to ensure that the individual V's are expressed in the same dimensionless variables because, if the dimensionless variables of each individual frequency are inserted straight into (5.71), they will be defined for each particular frequency. They will then not stand for the same real variables. If they are to be combined, they must be expressed by some definite variables - the same for all individual frequencies. Let us choose, (compare (2.73)- (2.77))

$$V^* = \frac{v}{v_0} \tag{5.197}$$

$$\theta^* = \omega_0 \tau \tag{5.198}$$

$$\sigma^* = \frac{\beta}{c_0^2}\omega_0 v_0 x \tag{5.199}$$

$$\epsilon^* = \frac{1}{2\beta}\frac{b\omega_0}{c_0 v_0 \rho_0} \tag{5.200}$$

Note that there appear two different b in this section. In (5.200) b is the dissipation parameter, all other b denotes a dimensionless frequency number. All single frequency boundary conditions written as

$$V^*(\sigma^* = 0, \theta^*) = a\sin(b\theta^* + \phi) = \frac{a}{2}[\exp(i(b\theta^* + \phi)) - \exp(-i(b\theta^* + \phi))] \ , \tag{5.201}$$

will yield results in the same dimensionless parameters by making the following replacements in (5.71)

$$V = \frac{V_*}{a} \tag{5.202}$$

$$\theta = b\theta^* + \phi \tag{5.203}$$

$$\sigma = ab\sigma^* \tag{5.204}$$

$$\epsilon = \frac{b}{a}\epsilon^* \tag{5.205}$$

So the solution to (2.76) with a single original frequency boundary condition $V_0 = a\sin(b\theta^*)$ will be, with the substitutions (5.202)-(5.205),

$$V^* = aV(\sigma^*, \theta^*)$$
$$= -\frac{4b\epsilon^* \sum_{n=1}^{\infty} \exp(-n^2b^2\epsilon^*\sigma^*) n(-1)^n I_n(\frac{a}{2b\epsilon^*}) \sin n(b\theta^* + \phi)}{I_0(\frac{a}{2b\epsilon^*}) + 2\sum_{n=1}^{\infty} \exp(-n^2b^2\epsilon^*\sigma^*)(-1)^n I_n(\frac{a}{2b\epsilon^*}) \cos n(b\theta^* + \phi)} . \tag{5.206}$$

Now we may write the superposition at zero propagated distance (5.195)

$$V^*_{mfq}(\sigma^* = 0, \theta^*) = 2i\epsilon^* \frac{\sum_{k=-\infty}^{\infty} kc_k e^{ik\theta^*}}{\sum_{k=-\infty}^{\infty} c_k e^{ik\theta^*}} \tag{5.207}$$

$$= \sum_{l=1}^{L} V_l^*(\sigma^* = 0, \theta^*) \tag{5.208}$$

$$\{ \text{ with (5.206) and introducing }_l = i(b_l\theta^* + \phi_l) \ \} \tag{5.209}$$

$$= -\sum_{l=1}^{L} \frac{\sum_{n=1}^{\infty} 2b_l\epsilon^* n(-1)^n I_n(\frac{a_l}{2b_l\epsilon^*})[e^{n\gamma_l} - e^{-n\gamma_l}]}{I_0(\frac{a_l}{2b_l\epsilon^*}) + \sum_{n=1}^{\infty}(-1)^n I_n(\frac{a_l}{2b_l\epsilon^*})[e^{n\gamma_l} + e^{-n\gamma_l}]} \tag{5.210}$$

$$= -\sum_{l=1}^{L} 2b_l\epsilon^* \frac{\sum_{n=-\infty}^{\infty} n(-1)^n I_n(\frac{a_l}{2b_l\epsilon^*}) \exp(in(b_l\theta^* + \phi_l))}{\sum_{n=-\infty}^{\infty}(-1)^n I_n(\frac{a_l}{2b_l\epsilon^*}) \exp(in(b_l\theta^* + \phi_l))} \tag{5.211}$$

$$= -\sum_{l=1}^{L} 2b_l\epsilon^* \frac{\sum_{n=-\infty}^{\infty} nA_n^{(l)} \exp(inb_l\theta^*)}{\sum_{n=-\infty}^{\infty} A_n^{(l)} \exp(inb_l\theta^*)} \tag{5.212}$$

with

$$A_n^{(l)} = I_n(\frac{a_l}{2b_l\epsilon^*})(-1)^n \exp(in\phi_l) \tag{5.213}$$

The relation between A_n and A_{-n} is

$$A_{-n}^{(l)} = A_n^{(l)} \exp(-2in\phi_l) \tag{5.214}$$

The next step is to identify (5.207) with (5.212) in each frequency with $n^{(l)}$ being the n belonging to the integer frequency number b_l

(and therefore also to $A_n^{(l)}$):

$$k = n^{(1)}b_1 + n^{(2)}b_2 + \ldots + n^{(L)}b_L = \sum_{l=1}^{L} n^{(l)}b_l \; . \tag{5.215}$$

As the numerator is simply the derivative of the denominator, the identification may be done in either numerator or denominator yielding

$$\sum_{k=-\infty}^{\infty} c_k e^{ik\theta^*} = \sum_{n^{(1)}=-\infty}^{\infty} A_n^{(1)} \exp(in^{(1)}b_1\theta^*) \sum_{n^{(2)}=-\infty}^{\infty} A_n^{(2)} \exp(in^{(2)}b_2\theta^*)$$

$$\ldots \sum_{n^{(L)}=-\infty}^{\infty} A_n^{(L)} \exp(in^{(L)}b_L\theta^*) \tag{5.216}$$

Through identification in frequency number k on the condition $k = \sum n^{(l)}b_l$, summarized over all n^l, the coefficients become

$$c_k = \sum_{n^{(1)}} \sum_{n^{(2)}} \cdots \sum_{n^{(L)}} A_{n^1}^{(1)} A_{n^2}^{(2)} \ldots A_{n^L}^{(L)} =$$

$$\sum_{n^{(1)}} \sum_{n^{(2)}} \cdots \sum_{n^{(L)}} I_{n^{(1)}}\left(\frac{a_1}{2\alpha_1\epsilon}\right)(-1)^{n^{(1)}} \exp(n^{(1)}\gamma_1)$$

$$\ldots I_{n^{(L)}}\left(\frac{a_L}{2\alpha_L\epsilon}\right)(-1)^{n^{(L)}} \exp(n^{(L)}\gamma_L) \; . \tag{5.217}$$

Equation (5.217), inserted in (5.195) is an exact analytical solution to Burgers' equation (2.76), valid for a boundary condition with any number of frequencies with arbitrary amplitudes and phases (Hedberg 1999).

This is an exact multifrequency solution in explicit form which includes dissipation. It is suitable as a replacement for numerical methods as it works fast on a computer. Once the coefficients c_k are obtained, the solution is is valid for all distances. Approximate analytical expressions describing the behaviour in limiting cases might be derived from this exact solution. But as the multifrequency condition contains many parameters – of which may be mentioned relative amplitudes, relative frequencies and relative phases – one can not expect to arrive at general asymptotic expressions.

When the dissipation ϵ^* gets smaller, a larger number of terms will have to be included in the series in (5.195). Due to the slow convergence of these series and the difference in size of the coefficients, there is a lower limit in ϵ below which the calculations lead to numbers exceeding the numerical representation in computers. This is the only limitation in the presented method – the same limitation as for the Cole-Mendousse solution (5.71).

5.5.2 Bifrequency solutions and creation of combination frequencies

The first example is $V_0 = \sin 7\theta + \sin 9\theta$ for the dissipation over non-linearity ratio $\epsilon^* = 0.05$ where the difference frequency, $9 - 7 = 2$, is not the same as the lowest frequency number b_0, which is always equal to 1. (All expressions are written in the normalized $\theta = \omega\tau$.)

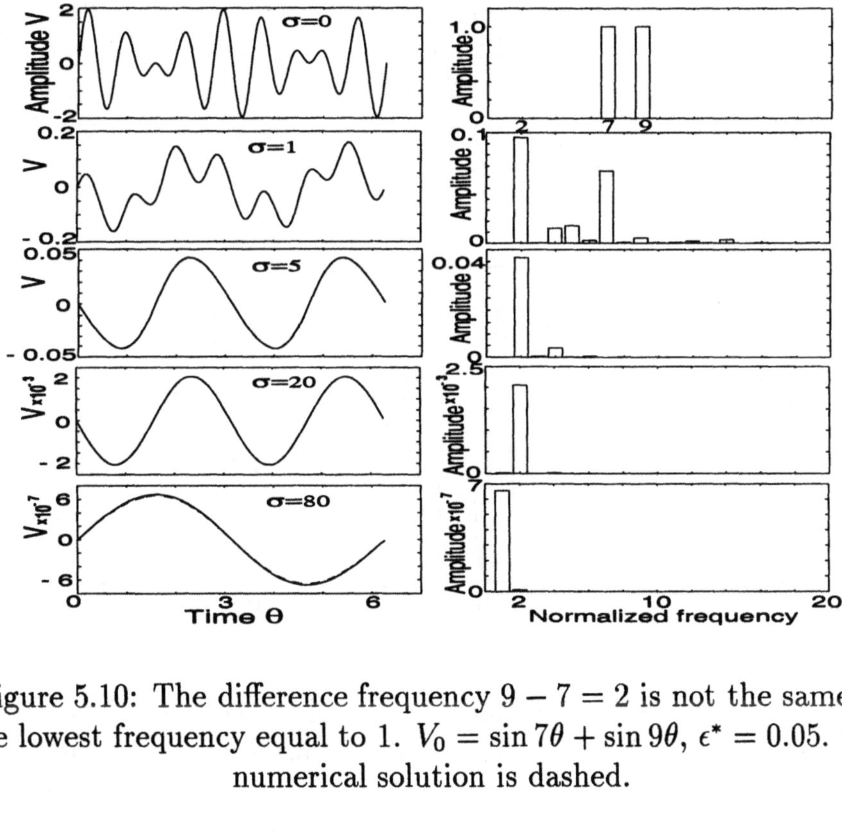

Figure 5.10: The difference frequency $9 - 7 = 2$ is not the same as the lowest frequency equal to 1. $V_0 = \sin 7\theta + \sin 9\theta$, $\epsilon^* = 0.05$. The numerical solution is dashed.

In Figure 5.10 is seen how combination frequencies are created. Note how the lowest frequency is not noticable as it can not be created directly from 7 and 9. Instead, it is created from e.g. $(4 \times 7 = 28) - (3 \times 9 = 27) = 1$, giving a very small contribution. At the distance $\sigma = 5$, the difference frequency has become dominant and many of the higher harmomics have been damped out. The wave stays in this approximate shape for a long distance, see for example $\sigma = 20$, so that one might believe this to be the final stage. But later on, at $\sigma = 80$, the higher damping of frequency 2 compared to that of frequency 1, has made the lowest frequency the only one remaining. The comparison with a numerical algorithm indicates the solution is correct.

To make a comparison with the Fenlon solution (Fenlon 1972) it is necessary to stay in the pre-shock region which in this case means up to the distance $\sigma_s \approx 1/(7+9) = 0.0625$. In Figure 5.11 is shown the multifrequency solution and the numerical solution for a dissipation of $\epsilon = 0.05$ and the Fenlon solution, which is inherently for zero dissipation, for the distances $\sigma = 0.02$ and 0.05. The present solution and the numerical solution coincide and their agreement with the Fenlon solution decreases with distance.

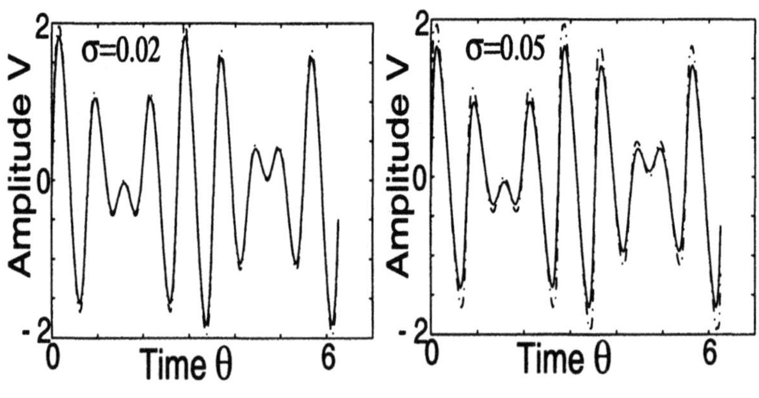

Figure 5.11: Hedberg's solution (1999) (solid), the numerical solution (dashed) and the Fenlon (1972) solution (dash-dotted) in the region of validity of the Fenlon solution (i.e. before shock formation), $V_0 = \sin 7\theta + \sin 9\theta$, $\epsilon^* = 0.05$.

This shows how pulses might also be calculated with multifrequency periodic methods. In Figure 5.12 is shown an N-wave pulse and its evolution with distance.

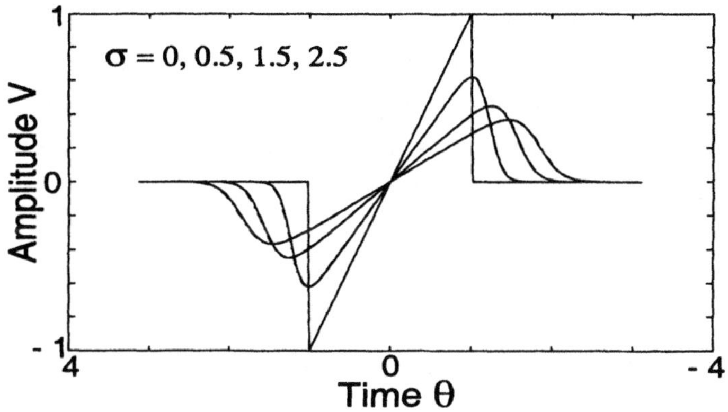

Figure 5.12: The initial N-wave and its evolution according to the multifrequency solution (solid), compared with a numerical algorithm (dash-dotted), $\epsilon^* = 0.04$.

An example is the specific situation when the difference frequency is the lowest frequency and also the same as one of the original frequencies: $V_0 = \sin(\theta + \phi) + \sin 2\theta$ with the dissipation $\epsilon = 0.05$. This can be compared to the the case of no dissipation, i.e. the Fenlon solution (4.101). Although one does not need to explicitly use the analytical expressions to calculate the numerical solutions, one might want to do that sometimes. We will in this case, as an example, give also the analytical specific expression for the coefficients c_k, found by specialization of (5.217):

$$c_k = \sum_{n=-\infty}^{\infty} (-1)^{k-n} e^{i(k-2n)\phi} I_{k-2n}(\frac{1}{2\epsilon}) I_n(\frac{1}{4\epsilon}) \tag{5.218}$$

In Figure 5.13 is seen the evolution for the phase $\phi = 0$ in the left column, and the phase $\phi = \pi/2$ in the right column. To begin with, at $\sigma = 0$, they of course have the same frequency content. At the distance $\sigma = 0.5$ is seen how the frequency 1 is larger for the phase $\phi = \pi/2$ and has increased its amplitude to above its original one. The frequency 2 on the other hand is larger for the phase $\phi = 0$, although it is not above its original amplitude. At a larger distance $\sigma = 40$, the frequency 1 is twice the size for the phase $\phi = \pi/2$ compared to the phase $\phi = 0$. Surprisingly enough, also frequency 2 is larger. It is approximately three times as large as for the phase $\phi = 0$. This is due to the nonlinear transfer of energy from frequency 1 to its first higher harmonic, frequency 2. For $\phi = 0$ the sharp shock created by nonlinearity has dissipated considerably more energy, than the smoother curve where $\phi = \pi/2$.

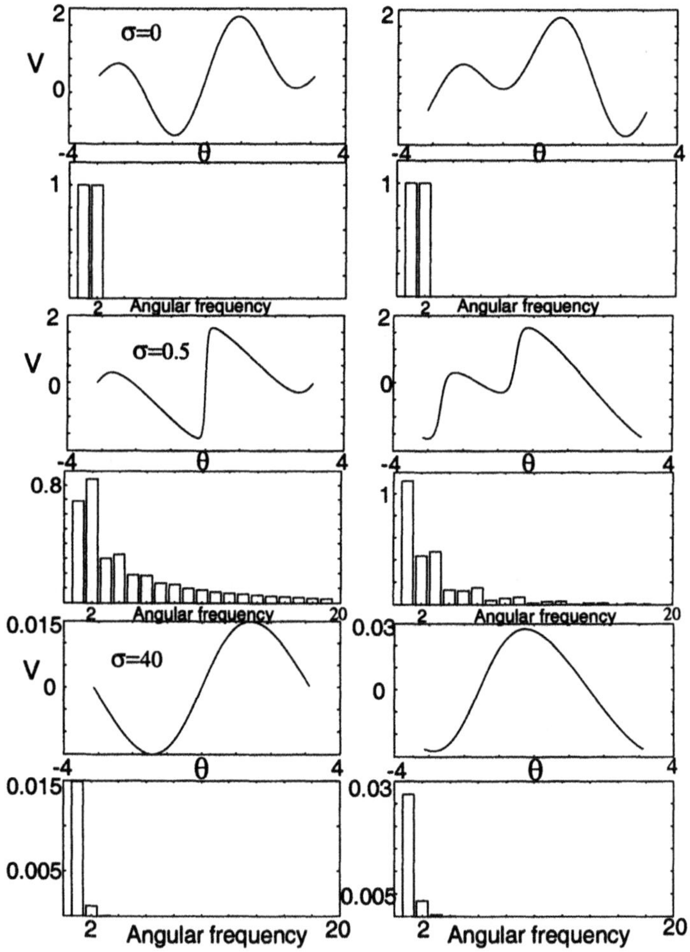

Figure 5.13: One of the original frequencies is the difference frequency and equal to the lowest frequency. $V_0 = \sin(\theta + \phi) + \sin 2\theta$, $\epsilon = 0.05$. Left: $\phi = 0$. Right: $\phi = \pi/2$.

In the next example, Figure 5.14, is shown a high frequency with small amplitude in the presence of a low frequency. The boundary condition is $V_0 = \sin\theta + 0.01 \sin 10\,\theta$. The oscillatory behaviour is seen not only for the frequency 10 but also for the surrounding components, in particular frequency 11 which has first one big and then one small oscillation. This disappears when the higher harmonics emanating from frequency 1 grows large enough to overwhelm the smaller contributions from the original frequency 10. Compare with Figure 4.16 for the lossless case.

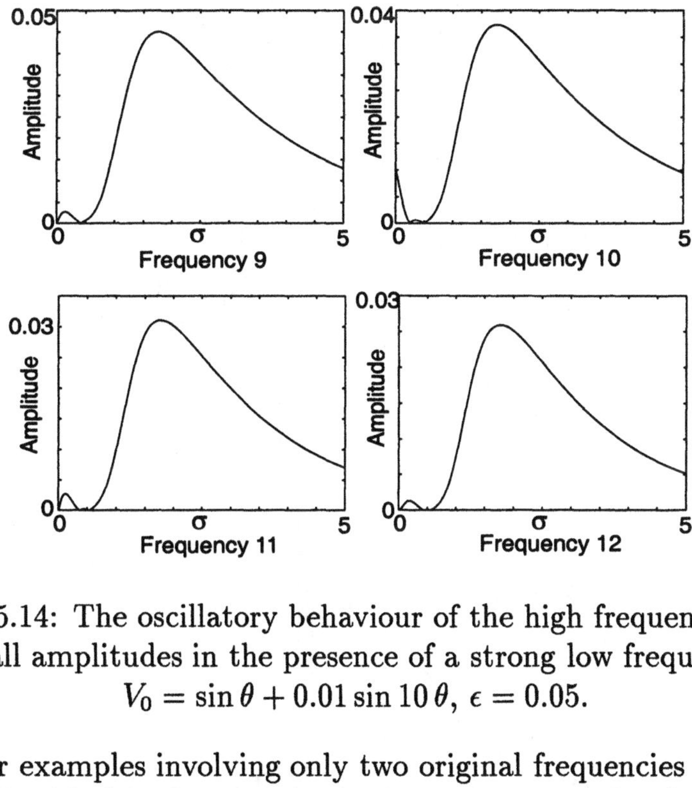

Figure 5.14: The oscillatory behaviour of the high frequencies with small amplitudes in the presence of a strong low frequency. $V_0 = \sin\theta + 0.01 \sin 10\,\theta$, $\epsilon = 0.05$.

So far examples involving only two original frequencies have been shown. In this fourth example the input is an amplitude-frequency modulated wave which in the nondissipative case has the property that, once the full shock structure has formed, the shocks will never merge but stand still and the frequency spectrum will have the same appearance while of course the amplitudes will decrease (Gurbatov and Hedberg 1998). The boundary condition in Figure 5.15 is $V_0 = (1 + 0.2 \cdot \cos\theta) \cdot \sin(10\theta + 10 \cdot 0.2 \sin\theta)$ with the dissipation over nonlinearity ratio $\epsilon^* = 0.002$.

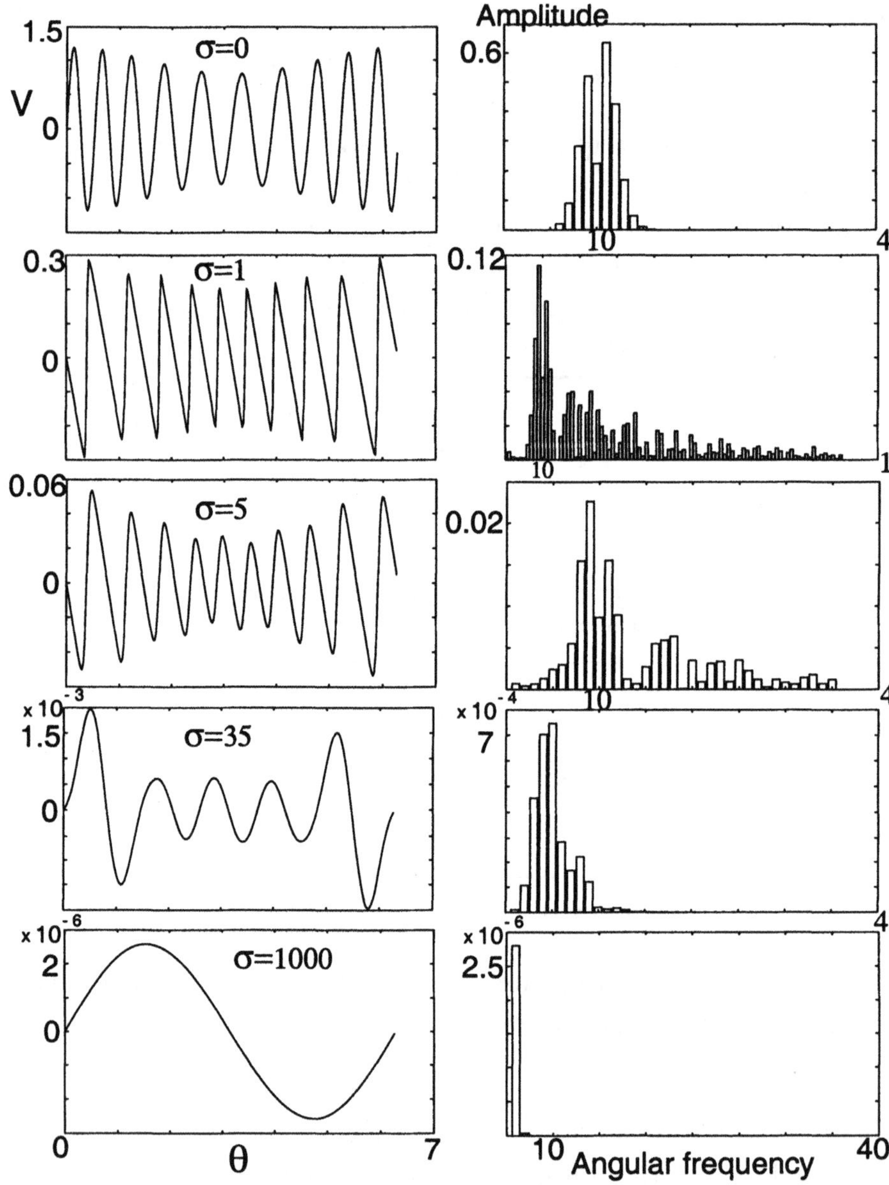

Figure 5.15: Evolution of an amplitude-frequency modulated wave. $V_0 = (1 + 0.2 * \cos\theta) \cdot \sin(10\theta + 10 \cdot 0.2 \sin\theta)$, $\epsilon^* = 0.002$

At $\sigma = 1$ in Figure 5.15 the look of the wave and its spectrum are almost exactly the same as for the case of no dissipation for all distances after the fully developed sawtooth shape - examplified by $\sigma = 5$ and $\sigma = 200$ in Figure 4.21. The spectrum contains a large number of higher harmonics. At $\sigma = 5$ the presence of damping smoothens the shocked triangular wave form and the spectrum is starting to change. At $\sigma = 35$ the effect is pronounced and the spectrum has only a small number of components. At $\sigma = 1000$ is shown how the lowest frequency present is the only one remaining, which is the final stage.

Investigations on bifrequency solutions of Burgers' equation using the saddle-point method on the Cole-Hopf solution were made by Cary (1973, 1975) and Coulouvrat (1991^c). Lardner (1982) extracted the old-age waveform studying the same problem by a generalization of Mendousse's (1953) solution (5.71). Fenlon (1974) derived solutions to Burgers' equation for the difference-frequency signal by a method of successive approximations.

Chapter 6

Nonlinear cylindrical and spherical diffusive waves

In this chapter we treat problems for which no exact solutions are known. Other methods than those used in the previous chapters must be used for the description of these physical phenomena. These new methods will not give analytical solutions valid in the entire domain of interest. Instead they give asymptotic waves, at large times after the start of the process, or for long distances from the boundary from which the wave is initiated. Dimensionless forms of generalized Burgers' equations for cylindrical and spherical waves are given. The method of finding the asymptotic form of a plane N-wave was given in section 5.1. This method is applied to cylindrical waves in this chapter. It also covers the calculation of the decay of the shock wave from a supersonic projectile, using asymptotics of the cylindrical N-waves. Finally the asymptotics of spherical and cylindrical periodic waves is studied.

Early studies on nonlinear cylindrical and spherical sound waves in a dissipative medium were made by Naugolnykh (1959) and Naugolnykh, Soluyan and Khokhlov (1962, 1963[a]).

Asymptotics of spherical pulses, without explicit reference to Burgers' equation, was studied by Naugolnykh (1973).

Nonlinear cylindrical and spherical waves are extensively studied in the book by Sachdev (1987).

6.1 Dimensionless generalized Burgers' equations

The case of a plane nonlinear acoustic wave is special in the sense that an exact solution of Burgers' equation describes it. In the cases of cylindrical and spherical nonlinear acoustic waves no exact solutions of the corresponding wave equations are known for travelling waves, and approximate and numerical methods must be used. As in the preceeding chapter, the wave equation written in physical variables has to be transformed to dimensionless form. The generalized Burgers' equations used for nonplanar waves may contain more than one dimensionless parameter. These dimensionless parameters have different meaning in the four cases treated in this chapter, namely cylindrical waves with initial and boundary conditions and spherical waves with initial and boundary conditions. We shall first derive non-dimensional forms of generalized Burgers' equations for these four cases bringing out distinctions therein.

We start with (2.87), rewritten in (6.1), which is a generalized Burgers' equation suitable for cylindrical (n=1) and spherical (n=2) waves generated at an initial time t_0:

$$\frac{\partial v}{\partial t} + \frac{n}{2t}v + \beta v \frac{\partial v}{\partial \rho} = \frac{b}{2\rho_0}\frac{\partial^2 v}{\partial \rho^2} \ . \tag{6.1}$$

Let l_0 be the length of the initial pulse or the wavelength of the initial wave, and let v_0 be the pulse height or the wave amplitude. We introduce the dimensionless variables V and ξ as (Crighton and Scott 1979):

$$V = \left(\frac{t}{t_0}\right)^{n/2} \frac{v}{v_0} \tag{6.2}$$

$$\xi = \frac{\rho}{l_0} \ . \tag{6.3}$$

The dimensionless parameter T_0 is defined as

$$T_0 = (\gamma + 1)\frac{v_0 t_0}{n l_0} \ . \tag{6.4}$$

In the cylindrical case a dimensionless time variable is defined as

$$T = 1 + (\gamma+1)v_0 t_0^{1/2}(t^{1/2} - t_0^{1/2})\frac{1}{l_0}\,. \tag{6.5}$$

Using (6.2)-(6.5) in (6.1) with n=1 we obtain the non-dimensional form of the equation

$$\frac{\partial V}{\partial T} + V\frac{\partial V}{\partial \xi} = \epsilon^{(1)}\frac{T - 1 + T_0}{2}\frac{\partial^2 V}{\partial \xi^2}\,, \tag{6.6}$$

where T_0 is given by (6.4) with $n = 1$ and $\epsilon^{(1)}$ is defined by

$$\epsilon^{(1)} = \frac{2b}{(\gamma+1)T_0\rho_0 v_0 l_0}\,. \tag{6.7}$$

For spherical waves the equations (6.2)-(6.4) are still used and a dimensionless time variable T is defined as

$$T = 1 + \frac{1}{2}(\gamma+1)\frac{v_0 t_0}{l_0}\ln\Big(\frac{t}{t_0}\Big)\,. \tag{6.8}$$

Using (6.2)-(6.4) and (6.8) in (6.1) with $n = 2$ we obtain

$$\frac{\partial V}{\partial T} + V\frac{\partial V}{\partial \xi} = \epsilon^{(2)}\exp\frac{T}{T_0}\frac{\partial^2 V}{\partial \xi^2}\,, \tag{6.9}$$

where T_0 is given by (6.4) with $n = 2$; here we have

$$\epsilon^{(2)} = \frac{b}{(\gamma+1)\rho_0 v_0 l_0}\exp\Big(-\frac{1}{T_0}\Big)\,. \tag{6.10}$$

The parameter T_0 defined by (6.4) can be chosen arbitrarily since t_0 is arbitrary and does not depend on the physical conditions. We may therefore choose T_0 equal to 1..

Two other generalized Burgers' equations will be derived from (2.85) for the cases where the cylindrical and spherical waves are generated from a boundary condition. In these cases the characteristic time of the problem is the inverse period or pulse duration, ω^{-1}. Now the dimensionless variables V and θ are introduced as

$$V = \Big(\frac{r}{r_0}\Big)^{\frac{n}{2}}\frac{v}{v_0} \tag{6.11}$$

$$\theta = \omega\tau\,, \tag{6.12}$$

where $n = 1, 2$ for cylindrical and spherical waves, respectively, r_0 is the radius of the source when the boundary condition is imposed, and v_0 is the amplitude of this disturbance. The dimensionless parameter R_0 is defined as

$$R_0 = (\gamma + 1)\frac{v_0\,\omega\,r_0}{2c_0^2} . \tag{6.13}$$

For cylindrical waves the dimensionless length variable R is defined as

$$R = R_0\Big(\big(\frac{r}{r_0}\big)^{\frac{1}{2}} - 1\Big) . \tag{6.14}$$

Using (6.11) - (6.14) in (2.85) with $n = 1$ we obtain (Cary 1967, 1968)

$$\frac{\partial V}{\partial R} - V\frac{\partial V}{\partial \theta} = \epsilon^{(3)}\big(\frac{R}{2} + R_0\big)\frac{\partial^2 V}{\partial \theta^2} , \tag{6.15}$$

where R_0 is given by (6.13) and $\epsilon^{(3)}$ as

$$\epsilon^{(3)} = \frac{b\omega}{(\gamma+1)R_0\rho_0 v_0 c_0} . \tag{6.16}$$

For spherical waves generated by boundary conditions, the length variable R is defined as

$$R = R_0 \ln\big(\frac{r}{r_0}\big) , \tag{6.17}$$

where R_0 is given by (6.13). Using (6.11)-(6.13) and (6.17) in (2.85) with $n = 2$ we obtain (Naugolnykh, Soluyan and Khokhlov 1963[a]; Cary 1967, 1968)

$$\frac{\partial V}{\partial R} - V\frac{\partial V}{\partial \theta} = \epsilon^{(4)} \exp\big(\frac{R}{R_0}\big)\frac{\partial^2 V}{\partial \theta^2} , \tag{6.18}$$

where R_0 is given by (6.13) and $\epsilon^{(4)}$ as

$$\epsilon^{(4)} = \frac{b\,\omega}{(\gamma+1)\rho_0 v_0 c_0} . \tag{6.19}$$

Cylindrical and spherical waves arising from boundary conditions are governed by (6.15) and (6.18) in a two-parameter space; R_0 depends on the physical conditions of the wave problem.

The equations (6.15) and (6.18) are used by Cary (1967, 1968) for numerical calculations of the distortion of cylindrical and spherical waves and by Fenlon (1973[b]) for a generalization of Banta's (1965) operational solution to the nonlinear wave equation for lossless fluids. An enlargement of (6.1) to other n values than 0, 1 and 2 is studied by Sachdev, Joseph and Nair (1994). A phenomenological approach to the problem of propagation of cylindrical and spherical nonlinear acoustic wave, adapted to the case in which, for example, the absorption depends on frequency, is given by Trivett and Van Buren (1981).

6.2 Cylindrical N-waves

6.2.1 Evolution of an initial cylindrical N-wave

As the first example of use of analytical methods in nonlinear wave problems lacking exact solutions we study the evolution of a cylindrical initial N-wave following Crighton and Scott (1979). The appropriate nonlinear wave equation in this case is given by (6.6). For later purposes we write a generalization of (6.6) and choose the notation ϵ for $\epsilon^{(1)}$:

$$\frac{\partial V}{\partial T} + V\frac{\partial V}{\partial \xi} = \epsilon\, g(T)\frac{\partial^2 V}{\partial \xi^2} \,. \tag{6.20}$$

The N-wave initial value problem is already solved in section 5.1 for (6.20) with $g(T)$ equal to 1. The calculation of V_0^* with arbitrary $g(T)$ and putting $C = 0$ gives, instead of (5.13),

$$V_0^* = \frac{1}{2}T^{-\frac{1}{2}}\left(1 - \tanh\frac{\xi^* - A(T)}{4g(T)T^{1/2}}\right) , \tag{6.21}$$

where $\xi^* = (\xi - T^{1/2})/\epsilon$ and $A(T)$ is found, in analogy with (5.27), as

$$A(T) = -T^{\frac{1}{2}}\int_1^T \frac{g(t)}{t}\, dt \,. \tag{6.22}$$

As mentioned in section 5.1 we may choose $T_0 = 1$ and so for the case of cylindrical wave, we have from (6.6)

$$g(T) = \frac{T}{2} \,. \tag{6.23}$$

In physical variables, the initial N-wave at time $t = t_0$ is

$$v(\rho, t_0) = v_0 \frac{\rho}{l_0} , \qquad |\rho| < l_0 \tag{6.24}$$

$$v(\rho, t_0) = 0 , \qquad |\rho| > l_0 . \tag{6.25}$$

The initial condition formulated in terms of dimensionless variables analogous to (4.27), (4.28) is

$$V(\xi, 1) = \xi , \qquad |\xi| < 1 \tag{6.26}$$

$$V(\xi, 1) = 0 , \qquad |\xi| > 1 . \tag{6.27}$$

As we know from (4.26), (4.27) the outer solution to (6.20) with the initial condition (6.26), (6.27) is

$$V_0(\xi, T) = \frac{\xi}{T} + o(\epsilon^n) , \qquad |\xi| < T^{\frac{1}{2}} \tag{6.28}$$

$$V_0(\xi, T) = 0 , \qquad |\xi| > T^{\frac{1}{2}} . \tag{6.29}$$

The lowest order inner solution V_0^* is found from (6.21) and (6.22) with $g(T) = T/2$:

$$V_0^*(\xi^*, T) = \frac{1}{2} T^{-\frac{1}{2}} \Big(1 - \tanh \frac{\xi^* - A(T)}{2T^{\frac{3}{2}}}\Big) \tag{6.30}$$

where

$$\xi^* = \frac{\xi - T^{\frac{1}{2}}}{\epsilon} \tag{6.31}$$

and

$$A(T) = -\frac{1}{2} T^{\frac{1}{2}} (T - 1) . \tag{6.32}$$

6.2.2 Four-step procedure for finding the asymptotic solution

The four-step procedure detailed in section 4.2 will now be used in finding the old-age solution from the cylindrical N-wave shock solution (Enflo 1998).

1) Solution in the fading shock region

In the shock region we use the variables (Y, T), where Y is defined as

$$Y = \frac{1}{\epsilon}\left(\frac{\xi}{T^{3/2}} - \frac{1+\epsilon/2}{T}\right) . \tag{6.33}$$

The shock tail is defined by the relations

$$\epsilon T \ll 1 \tag{6.34}$$

$$Y > \ln\left(\frac{1}{\delta}\right), \qquad \delta \ll 1 . \tag{6.35}$$

The solution in the shock tail region is obtained from (6.30), using (6.33) and (6.35)

$$V(Y,T) = T^{-\frac{1}{2}} \exp\left(-\frac{1}{2} - Y\right)(1 + O(\delta) + O(\epsilon)) . \tag{6.36}$$

The term $O(\delta)$ in (6.36) signifies that V_0^* in (6.30) is approximated under the condition (6.35). The term $O(\epsilon)$ in (6.36) arises from the fact that V_0^* is the first term in an asymptotic expansion of the shock solution in powers of ϵ.

The condition for (6.36) to be the solution of a linear equation will now be determined. Using the scaling

$$W(Y,T) = \frac{V}{\delta} \tag{6.37}$$

and the variable Y from (6.33), instead of ξ, in (6.20) and (6.23), we obtain

$$\epsilon\left(\frac{\partial W}{\partial T} - \frac{3Y}{2T}\frac{\partial W}{\partial Y}\right) + \frac{\delta}{T^{\frac{3}{2}}} W \frac{\partial W}{\partial Y} - \frac{1}{2T^2}\frac{\partial W}{\partial Y} = \frac{1}{2T^2}\frac{\partial^2 W}{\partial Y^2} . \tag{6.38}$$

From (6.35) it is obvious that the variable Y should be scaled by $\ln(\frac{1}{\delta})$. Taking this fact into account we find that the nonlinear term in (6.38) is of the order $\delta/\ln(1/\delta)$. In order to neglect this term in (6.38) we have

$$\frac{\delta}{\ln(1/\delta)} \ll \epsilon \ll \frac{1}{(\ln(1/\delta))^2} . \tag{6.39}$$

The details of the present calculation are given only in the lowest order in ϵ. The first inequality in (6.39) is essential if we consider terms of the order ϵ in (6.36). It is therefore seen that the first term in (6.36) solves (6.38) if we neglect order δ- and ϵ-terms.

2) Integral representation of the solution in the old-age region

In the old-age regime, where $T > 1/\epsilon$, we introduce the scaling

$$T' = \epsilon T \ , \qquad \xi' = \epsilon^{\frac{1}{2}}\xi \ , \qquad V' = \epsilon^{-\frac{1}{2}}V \tag{6.40}$$

and obtain from (6.20) and (6.23),

$$\frac{\partial V'}{\partial T'} + V'\frac{\partial V'}{\partial \xi'} = \frac{T'}{2}\frac{\partial^2 V'}{\partial \xi'^2} \ . \tag{6.41}$$

For large T' values the nonlinear term in (6.41) can be neglected. (This fact will be verified after solving (6.41) with the nonlinear term neglected.) The general solution of the linear equation,

$$\frac{\partial V'}{\partial T'} = \frac{T'}{2}\frac{\partial^2 V'}{\partial \xi'^2} \ , \tag{6.42}$$

can be represented as an integral, analogous to (5.49):

$$V'(\xi', T') = \int_0^\infty \exp(-\lambda^2 T'^2 + 2i\lambda\xi')h(\lambda)\, d\lambda + c.c. \ , \tag{6.43}$$

where *c.c.* refers to the complex conjugate. By assuming the lower limit of the integral in (6.43) to be 0 and not $-\infty$ we anticipate the singularity of $h(\lambda)$ at $\lambda = 0$.

3) Determination of the old-age integral representation by evaluating it in the fading shock region

The shock tail variables (ξ, Y) will now be used in the integral representation (6.43); also a change of variable will be made to transform the integral in (6.43) into an integral suitable for evaluation in the shock tail region by the steepest descent method (see subsection 3.2.2). After the change of variable

$$\lambda = \epsilon^{-3/2}\kappa \ , \tag{6.44}$$

the substitution

$$h(\lambda) = \epsilon^{3/2}g(\kappa) \ , \tag{6.45}$$

and change of (ξ', T') to (ξ, T) by means of (6.40) and (6.33), we obtain

$$V' = \int_0^\infty \exp\{\frac{1}{\epsilon}[-\kappa^2 T^2 + 2i\kappa(T^{1/2} + \epsilon Y T^{3/2} + \frac{\epsilon}{2} T^{1/2})]\} g(\kappa) d\kappa + \text{c.c.} \; . \tag{6.46}$$

The substitution (6.44) thus gives two terms of the order ϵ^{-1} balancing each other in the argument of the exponential function on the righthand side of (6.46). Because of the smallness of ϵ the integral in (6.46) is suitable for evaluation by the steepest descent method.

In order to contribute significantly to the steepest descent integral the function $g(\kappa)$ must be the product of a slowly varying function and an exponential function, whose exponent is of the order ϵ^{-1}. We attempt the following expression for $g(\kappa)$:

$$g(\kappa) = G(\kappa) \exp(\gamma \frac{\kappa^s}{\epsilon}), \tag{6.47}$$

where $G(\kappa)$ is a slowly varying function of κ and γ is a constant. Written as a function of λ the exponent in (6.47) must not contain ϵ. This condition and (6.44) give

$$\lambda^s \frac{\epsilon^{\frac{3}{2}s}}{\epsilon} = \lambda^s \tag{6.48}$$

and so

$$s = \frac{2}{3} \; . \tag{6.49}$$

Equations (6.47) and (6.49) change (6.46) to

$$V' = \int_0^\infty \exp[\frac{1}{\epsilon} \chi(\kappa, Y, T)] G(\kappa) \, d\kappa + \text{c.c.} \; , \tag{6.50}$$

where

$$\chi(\kappa, Y, T) = -\kappa^2 T^2 + 2i\kappa T^{\frac{1}{2}} + \gamma \kappa^{\frac{2}{3}} + 2i\kappa\epsilon(Y T^{\frac{3}{2}} + \frac{1}{2} T^{\frac{1}{2}}) \; . \tag{6.51}$$

We observe that the term $\gamma\kappa^{\frac{2}{3}}$ in $\chi(\kappa, Y, T)$ makes the integrand in (6.50) singular at $\kappa = 0$. Therefore we must choose 0 and not $-\infty$ as the lower limit of the integral in (6.50).

The integral (6.50) is evaluated by the steepest descent method, giving the following result in the lowest approximation:

$$V = \epsilon^{1/2} V' = \pi^{1/2} \exp\Big(\frac{\chi(\kappa_0)}{\epsilon}\Big)\Big(-\frac{2}{\chi''(\kappa_0)}\Big)^{1/2} G(\kappa_0) + c.c. \,, \qquad (6.52)$$

where κ_0 is the saddle-point

$$\kappa_0 = \kappa_0^{(0)} + \epsilon\kappa_0^{(1)} + O(\epsilon^2) \,. \qquad (6.53)$$

The expression (6.52) along with (6.53) will now be identified with the expression (6.36). Because terms of the order ϵ^{-1} do not occur in the argument of the exponential in (6.36) we must require

$$\chi(\kappa_0) = O(\epsilon) \,. \qquad (6.54)$$

Equation (6.54) shows that there must be terms of order ϵ in the expansion (6.53). Furthermore, we must require that at the saddle-point $\kappa = \kappa_0$, $\chi'(\kappa_0)$ is one order in ϵ smaller than $\chi(\kappa_0)$, i.e.

$$\chi'(\kappa_0) = O(\epsilon^2) \,. \qquad (6.55)$$

Equation (6.54) gives one equation and (6.55) gives (in orders ϵ^0 and ϵ^1) two equations for solving the unknowns $\kappa_0^{(0)}$, $\kappa_0^{(1)}$ and γ. From (6.54) and (6.51) we obtain

$$-\kappa_0^{(0)^2} T^2 + 2i\kappa_0^{(0)} T^{\frac{1}{2}} + \gamma{\kappa_0}^{(0)\frac{2}{3}} = 0 \,. \qquad (6.56)$$

Equation (6.55), together with (6.51), gives

$$-2\kappa_0^{(0)} T^2 + 2iT^{\frac{1}{2}} + \frac{2}{3}\gamma{\kappa_0}^{(0)-\frac{1}{3}} = 0 \qquad (6.57)$$

$$-2\kappa_0^{(1)} T^2 + 2i\kappa_0^{(0)}(YT^{\frac{3}{2}} + \frac{1}{2}T^{\frac{1}{2}}) - \frac{2}{9}\gamma\kappa_0^{(0)-\frac{4}{3}}\kappa_0^{(1)} = 0 \,. \qquad (6.58)$$

From (6.56) and (6.57) we obtain

$$\kappa_0^{(0)} = \frac{i}{2} T^{-\frac{3}{2}} \qquad (6.59)$$

$$\gamma\kappa_0^{(0)-\frac{1}{3}} = -\frac{3}{2} i T^{\frac{1}{2}} \,. \qquad (6.60)$$

Using (6.59) and (6.60) in (6.58) we obtain

$$\kappa_0^{(1)} = \frac{3}{2} i (Y T^{-\frac{1}{2}} + \frac{1}{2} T^{-\frac{3}{2}}) \; . \tag{6.61}$$

Equations (6.59) - (6.61), (6.53) and (6.51) give

$$\chi(\kappa_0) = -\epsilon (Y + \frac{1}{2} T^{-1}) + O(\epsilon^2) \tag{6.62}$$

$$\chi''(\kappa_0) = -\frac{4}{3} T^2 + O(\epsilon) \; . \tag{6.63}$$

We will now equate the steepest descent solution (6.52), with $\chi(\kappa_0)$ and $\chi''(\kappa_0)$ inserted from (6.62) and (6.63) and the shock tail solution (6.36). The lowest order in ϵ calculation implies that $\kappa_0^{(0)}$ from (6.59) is inserted as the argument of $G(\kappa_0)$ in (6.52) [1]:

$$G(\kappa_0^{(0)}) + G^*(\kappa_0^{(0)*}) = \left(\frac{2}{3\mathrm{e}}\right)^{\frac{1}{2}} \pi^{-\frac{1}{2}} T^{\frac{1}{2}} \exp\left(\frac{1}{2T}\right) . \tag{6.64}$$

Choosing $G(\kappa_0^{(0)})$ real and using (6.59) we obtain

$$G(\kappa) = \frac{2^{-1/3} \pi^{-1/2} \mathrm{e}^{i\pi/6}}{\sqrt{6\mathrm{e}}} \kappa^{-\frac{1}{3}} \exp(2^{-\frac{1}{3}} \mathrm{e}^{-i\frac{\pi}{3}} \kappa^{\frac{2}{3}}) \; . \tag{6.65}$$

In (6.65) we have used the following solution (there are three possible third roots of i) of (6.59):

$$T^{\frac{1}{2}} = 2^{-\frac{1}{3}} \mathrm{e}^{i\frac{\pi}{6}} \kappa_0^{(0)-\frac{1}{3}} \; . \tag{6.66}$$

In order that (6.66) be consistent with (6.60) we must have

$$\gamma = 2^{-\frac{1}{3}} \frac{3}{2} \mathrm{e}^{-i\frac{\pi}{3}} \; . \tag{6.67}$$

Using (6.44), (6.33) and (6.40) we write (6.51) as

$$\frac{1}{\epsilon} \chi(\lambda, Y, T) = -\lambda^2 T'^2 + 2i\lambda\xi' + \gamma\lambda^{\frac{2}{3}} \; . \tag{6.68}$$

[1]Here * indicates the complex conjugate.

With (6.44) and (6.65) we write (6.50) as an integral representation of the old-age solution with neglect of higher orders of ϵ than ϵ^0:

$$\begin{aligned} V' &= \frac{V}{\sqrt{\epsilon}} \\ &= \frac{2^{-\frac{1}{3}}\pi^{-\frac{1}{2}}e^{i\frac{\pi}{6}}}{\sqrt{6e}} \int_0^\infty \exp(-\lambda^2 T'^2 + 2i\lambda\xi' + \gamma\lambda^{\frac{2}{3}})\lambda^{-\frac{1}{3}}\, d\lambda + c.c. \\ &= \frac{2^{-\frac{1}{3}}\pi^{-\frac{1}{2}}}{\sqrt{6e}}(I_1 + I_1^*)\ , \end{aligned} \tag{6.69}$$

with the introduction of the notation I_1 for the relevant integral. The integral (6.69) is a solution of the linear equation (6.42) having the required lowest order behaviour in the fading shock region. However, because of the singularity of the integrand at $\lambda = 0$, the integral expression (6.69) is ambiguous and a choice must be made of the appropriate Riemann sheet for its evaluation. This matter will now be dealt with.

4) Evaluation of the amplitude of the old-age solution

We first make a straightforward evaluation of the expression (6.69), using the integral representation of Hermite functions (5.61). The series expansion of exponential function $\exp(\gamma\lambda^{\frac{2}{3}})$ in (6.69),

$$\exp(\gamma\lambda^{\frac{2}{3}}) = 1 + \gamma\lambda^{\frac{2}{3}} + \frac{1}{2!}\gamma^2\lambda^{\frac{4}{3}} + \frac{1}{3!}\gamma^3\lambda^2 + \dots\ , \tag{6.70}$$

is first inserted into (6.69) and the integral is evaluated term by term. We obtain

$$\begin{aligned} I_1 + I_1^* = \pi^{\frac{1}{2}} \exp(-\frac{\xi'^2}{T'^2})[&\frac{2^{\frac{1}{3}}}{T'^{\frac{2}{3}}} H_{-\frac{1}{3}}(\frac{\xi'}{T'}) \\ &+ \frac{2^{-\frac{1}{3}}}{T'^{\frac{4}{3}}}\gamma e^{i\frac{\pi}{3}} H_{\frac{1}{3}}(\frac{\xi'}{T'}) + \frac{1}{2!}\frac{2^{-1}}{T'^2}\gamma^2 e^{i\frac{2\pi}{3}} H_1(\frac{\xi'}{T'}) + \dots]\ . \end{aligned} \tag{6.71}$$

The integral in (6.69) is evaluated on that Riemann sheet in the complex λ plane, where $\lambda^{\frac{2}{3}} = |\lambda|^{\frac{2}{3}}$ for λ real and positive. However, it is immediately clear that the series on the righthand side of (6.71), although behaving correctly in the fading shock region, is not the correct approximate solution of the initial value problem (6.20), (6.23), (6.26), (6.27). The shock solution (6.28)-(6.32) fulfils the following conditions, which we also require the old-age solution to satisfy:

a) The solution is an entire function of ξ and is zero at $\xi = 0$.

b) The solution decreases faster than every negative power of ξ for $\xi \to \pm\infty$ for fixed T.

In order that the nonlinear term in (6.41) be negligible for large T' we require that

c) The solution decreases faster than T'^{-1} for $T' \to \infty$ with ξ' fixed.

The series expansion of the Hermite functions $H_\nu(z)$ around $z = 0$ is (Lebedev 1965):

$$H_\nu(z) = \frac{1}{2\Gamma(-\nu)} \sum_{m=0}^{\infty} \frac{(-1)^m \Gamma(\frac{m-\nu}{2})}{m!} (2z)^m \,, \qquad |z| < \infty \,. \tag{6.72}$$

It is obvious from (6.72) that the Hermite functions of noninteger order in (6.71) do not fulfil the requirement a) above. The first term of the series in (6.71) does not fulfil requirement c). If we, with Lebedev (1965), agree that the indeterminate (of the form $\frac{\infty}{\infty}$) ratio $\frac{\Gamma(-1)}{\Gamma(-2)}$ is formally equal to -4, the formula (6.72) can be used for nonnegative integral $\nu = n$ if all indeterminate expressions are evaluated accordingly. Then it is seen from (6.72) that the terms with Hermite functions of positive odd-integer order in (6.71) fulfil the requirement a) above.

For large $|z|$ the asymptotic representation of $H_\nu(z)$ is (Lebedev 1965):

$$H_\nu(z) = (2z)^\nu \{ \sum_{k=0}^{n} \frac{(-1)^k}{k!} (-\nu)_{2k} (2z)^{-2k} + O(|z|^{-2n-2}) \} \,,$$

$$|\arg z| \le \frac{3\pi}{4} - \delta \tag{6.73}$$

$$H_\nu(z) = (2z)^\nu \{ \sum_{k=0}^{n} \frac{(-1)^k}{k!} (-\nu)_{2k} (2z)^{-2k} + O(|z|^{-2n-2}) \}$$
$$- \frac{\sqrt{\pi} e^{\nu\pi i}}{\Gamma(-\nu)} e^{z^2} z^{-\nu-1} \{ \sum_{k=0}^{n} \frac{(\nu+1)_{2k}}{k!} (2z)^{-2k} + O(|z|^{-2n-2}) \} \,,$$

$$\frac{\pi}{4} + \delta \leq \arg z \leq \frac{5\pi}{4} - \delta\,. \qquad (6.74)$$

From (6.74) it is seen that the requirement b) is not fulfilled for $\xi' \to -\infty$ by the terms in (6.71) with Hermite functions of noninteger order. In the series on the righthand side of (6.71) the terms with Hermite functions of odd-integer order, i.e. Hermite polynomials, fulfil the requirements a) - c) above. The question then arises: does another way of evaluating the integral give a series expansion of the solution where only odd-integer Hermite functions occur ? We recollect that the integral in (6.71) is evaluated on one of three possible Riemann sheets. Now we study the same integral evaluated on each of the three sheets. The integrals I_1,I_2, ... I_6 are defined through

$$I_{1,3,5} = \int_0^\infty \exp(-\lambda^2 T'^2 + 2i\lambda\xi' + \gamma\lambda^{\frac{2}{3}} + i\frac{\pi}{6})\lambda^{-\frac{2}{3}}\, d\lambda \qquad (6.75)$$

with

$$\lambda^{\frac{2}{3}} = |\lambda|^{\frac{2}{3}} \quad \text{in} \quad I_1$$
$$\lambda^{\frac{2}{3}} = |\lambda|^{\frac{2}{3}} e^{i\frac{2\pi}{3}} \quad \text{in} \quad I_3$$
$$\lambda^{\frac{2}{3}} = |\lambda|^{\frac{2}{3}} e^{-i\frac{2\pi}{3}} \quad \text{in} \quad I_5$$

and

$$I_{2,4,6} = \int_0^\infty \exp(-\lambda^2 T'^2 - 2i\lambda\xi' + \gamma\lambda^{\frac{2}{3}} + i\frac{\pi}{6})\lambda^{-\frac{2}{3}} d\lambda \qquad (6.76)$$

with

$$\lambda^{\frac{2}{3}} = |\lambda|^{\frac{2}{3}} \quad \text{in} \quad I_2$$
$$\lambda^{\frac{2}{3}} = |\lambda|^{\frac{2}{3}} e^{i\frac{2\pi}{3}} \quad \text{in} \quad I_4$$
$$\lambda^{\frac{2}{3}} = |\lambda|^{\frac{2}{3}} e^{-i\frac{2\pi}{3}} \quad \text{in} \quad I_6\,.$$

The number γ in (6.75) and (6.76) is given by (6.67). From (6.75) and (6.76) it is easily seen that

$$I_4 = -I_1^* \qquad (6.77)$$

$$I_2 = -I_3^* \tag{6.78}$$

The integral I_3 is now evaluated in the shock tail region. After a change of variables according to (6.44), (6.33), (6.40) we obtain from (6.75),

$$I_3 = \epsilon^{-1} e^{i\frac{2\pi}{3}} \int_0^\infty \exp\{\frac{1}{\epsilon}[-\kappa^2 T^2 + 2i\kappa(T^{\frac{1}{2}} + \epsilon Y T^{\frac{2}{3}}) + \gamma^* \kappa^{\frac{2}{3}}]\} e^{i\frac{\pi}{6}} \kappa^{-\frac{1}{3}} \, d\kappa \, . \tag{6.79}$$

Putting

$$\eta = \kappa_0^{(0)\frac{1}{3}} T^{\frac{1}{2}} \tag{6.80}$$

we obtain from (6.79) the following equation for the saddle-point in the zeroth approximation

$$-2\eta^4 + 2i\eta + 2^{-\frac{1}{3}} e^{i\frac{\pi}{3}} = 0 \, . \tag{6.81}$$

We are now interested in a solution of (6.81) lying on that Riemann sheet in the complex κ-plane on which the integral (6.79) is evaluated. On this Riemann sheet $\kappa^{-\frac{1}{3}}$ is real and positive when κ is real and positive. This means that the required solution for $\kappa^{(0)\frac{1}{3}}$ and hence for η must have a phase angle between $-\frac{\pi}{3}$ and $\frac{\pi}{3}$. Excluding the root

$$\eta = 2^{-\frac{1}{3}} e^{i\frac{5\pi}{6}} \, , \tag{6.82}$$

which does not fulfil this condition, and putting

$$\eta = 2^{-\frac{1}{3}} e^{-i\frac{\pi}{6}} \zeta \, , \tag{6.83}$$

the remaining three roots are given by the equation

$$\zeta^3 - \zeta^2 + \zeta + 1 = 0 \, . \tag{6.84}$$

The two complex roots of (6.84) give values of $\kappa_0^{(0)\frac{1}{3}}$ lying on the relevant Riemann sheet. For both these values of $\kappa_0^{(0)\frac{1}{3}}$ we find that the real part of the exponent in the integrand of (6.79) is negative and of the order ϵ^{-1}. Thus the integrand I_3 given by (6.79) has a vanishing value in the shock tail region.

In the same manner we find that the integrals I_5 and I_6 have vanishing values in the shock tail region. Because of their vanishing in the fading shock, the integrals I_3, I_5 and I_6 can be used for changing the unsatisfactory behaviour of the series expansion (6.71) of the integral in the old-age solution (6.69). A series expansion of I_5 similar to that for $I_1 + I_1^*$ in (6.71) gives, on using (6.67),

$$\begin{aligned} I_5 &= -\frac{i}{T'^{\frac{2}{3}}}[\Gamma(\frac{2}{3})H_{-\frac{2}{3}}(-i\frac{\xi'}{T'}) - \frac{3}{2^{\frac{4}{3}}}\Gamma(\frac{4}{3})H_{-\frac{4}{3}}(-i\frac{\xi'}{T'})\frac{1}{T'^{\frac{2}{3}}} \\ &+ \frac{1}{2!}\frac{3^2}{2^{\frac{8}{3}}}\Gamma(2)H_{-2}(-i\frac{\xi'}{T'})\frac{1}{T'^{\frac{4}{3}}} - \frac{1}{3!}\frac{3^3}{2^4}\Gamma(\frac{8}{3})H_{-\frac{8}{3}}(-i\frac{\xi'}{T'})\frac{1}{T'^2} + \ldots] \end{aligned} \tag{6.85}$$

where we have used the relation (Lebedev 1965)

$$H_{-\nu}(z) = \frac{1}{\Gamma(\nu)}\int_0^\infty \exp(-t^2 - 2tz)t^{\nu-1}\,dt\ , \quad \mathrm{Re}\{\nu\} > 0\ . \tag{6.86}$$

The Hermite functions of negative order in (6.85) are replaced by Hermite functions of positive order (Lebedev 1965):

$$\begin{aligned} &H_{-\nu}(-iz) \\ &= \frac{\Gamma(-\nu+1)}{2^\nu \pi^{\frac{1}{2}}}\exp(-z^2)\exp(-i\frac{\nu\pi}{2})[H_{\nu-1}(z) - \mathrm{e}^{i(\nu-1)\pi}H_{\nu-1}(-z)]\ . \end{aligned} \tag{6.87}$$

Using (6.76), (6.85), the formula

$$\Gamma(z)\Gamma(1-z) = \frac{\pi}{\sin \pi z}\ , \tag{6.88}$$

and (Lebedev 1965)

$$H_\nu(z) = \frac{2^\nu\Gamma(\nu+1)}{\pi^{\frac{1}{2}}}\mathrm{e}^{z^2}[\mathrm{e}^{i\frac{\pi\nu}{2}}H_{-\nu-1}(iz) + \mathrm{e}^{-i\frac{\pi\nu}{2}}H_{-\nu-1}(-iz)]\ , \tag{6.89}$$

we obtain

$$\begin{aligned} I_5 - I_6 &= -2^{\frac{1}{3}}\pi^{\frac{1}{2}}\exp(-\frac{\xi'^2}{T'^2})\{\frac{1}{T'^{\frac{2}{3}}}[H_{-\frac{1}{3}}(\frac{\xi'}{T'}) - H_{-\frac{1}{3}}(-\frac{\xi'}{T'})] \\ &+ \frac{3}{4}\frac{1}{T'^{\frac{4}{3}}}[H_{\frac{1}{3}}(\frac{\xi'}{T'}) - H_{\frac{1}{3}}(-\frac{\xi'}{T'})] + \frac{1}{2!}\left(\frac{3}{4}\right)^2\frac{1}{T'^2}H_1(\frac{\xi'}{T'})\ldots\} \end{aligned} \tag{6.90}$$

Using (6.75) and (6.76) we get

$$I_1 - I_4 = 2^{\frac{1}{3}}\pi^{\frac{1}{2}} \exp(-\frac{\xi'^2}{T'^2})[\frac{1}{T'^{\frac{2}{3}}} H_{-\frac{1}{3}}(\frac{\xi'}{T'}) + \frac{3}{4}\frac{1}{T'^{\frac{4}{3}}} H_{\frac{1}{3}}(\frac{\xi'}{T'}) + \frac{1}{2!}\Big(\frac{3}{4}\Big)^2 \frac{1}{T'^2} H_1(\frac{\xi'}{T'}) \ldots] \, . \quad (6.91)$$

Using (6.75)-(6.78), $I_3 - I_2$ is obtained from $I_1 - I_4$ by changing ξ' into $-\xi'$ and changing the sign. We find that in the sum $I_1 - I_2 + I_3 - I_4 + I_5 - I_6$, all Hermite functions of non-integral order cancel and Hermite polynomials of odd-integral order appear, each three times:

$$I_1 - I_2 + I_3 - I_4 + I_5 - I_6 = 3 \cdot 2^{\frac{1}{3}} \cdot \pi^{\frac{1}{2}} \exp(-\frac{\xi'^2}{T'^2}) \sum_{k=0}^{\infty} \frac{1}{(3k+2)!} (\frac{3}{4})^{3k+2} \frac{1}{T'^{2k+2}} H_{2k+1}(\frac{\xi'}{T'}) \, . \quad (6.92)$$

The sum at the lefthand side of (6.92) can be represented as a line integral:

$$I_1 - I_2 + I_3 - I_4 + I_5 - I_6 = \int_{C_1+C_2+C_3} [\exp(-\lambda^2 T'^2 + 2i\lambda\xi' + \gamma\lambda^{\frac{2}{3}} + i\frac{\pi}{6}) - \exp(-\lambda^2 T'^2 - 2i\lambda\xi' + \gamma\lambda^{\frac{2}{3}} + i\frac{\pi}{6})]\lambda^{-\frac{1}{3}} \, d\lambda \, , \quad (6.93)$$

where C_1, C_2 and C_3 are the real axis from 0 to ∞ on the three Riemann sheets, respectively, of the analytic function $\lambda^{\frac{1}{3}}$, which is chosen to be unique in a plane cut from $-\infty$ to 0. Our old-age solution (6.69), with $I_1 + I_1^*$ expanded according to (6.71), thus must be replaced by a series containing only Hermite functions of integral order multiplied by 3. We write down the first term of this series as *

$$V(\xi', T') = \epsilon^{\frac{1}{2}} \frac{C}{2T'^2} \exp(-\frac{\xi'^2}{T'}) H_1(\frac{\xi'}{T'}) + O(\epsilon) = \epsilon^{\frac{1}{2}} C \frac{\xi'}{T'^3} \exp(-\frac{\xi'^2}{T'}) + O(\epsilon) \quad (6.94)$$

where the old-age constant is given as

$$C = \frac{27}{16} \frac{1}{\sqrt{6e}} \approx 0.4178 \, . \quad (6.95)$$

Cylindrical N-waves were first studied numerically by Sachdev and Seebass (1973). Numerical calculations of C has been made (Sachdev, Tikekar and Nair 1986; Hammerton and Crighton 1989) in which the constant C is found to be equal to 0.34 or 0.35. If the shock structure solution (6.30) is expanded to the order ϵ (Crighton and Scott 1979), corrections to C both of the order ϵ^0 and ϵ are obtained. Such a calculation was made by Enflo (1998) with the result:

$$C = \frac{27}{16}\frac{1}{\sqrt{6e}}[1+2\epsilon-\frac{1}{2}(1+\frac{8}{3}\epsilon)(\frac{\pi^2}{6}-\ln 2-\frac{11}{36})] = 0.2828+O(\epsilon) \ . \quad (6.96)$$

The value 0.34 of the old-age constant was found for very small ϵ. However, for $\epsilon = \frac{1}{10}$, we expect a 20 percent enhancement of C by using (6.96). This enhancement was found numerically by Hammerton and Crighton (1989), who found $C = 0.43$ for $\epsilon = \frac{1}{10}$, and $C = 0.35$ for $\epsilon = \frac{1}{1000}$.

Thus the results of the present calculation of C are compatible with the results of numerical calculations. The principal gains of the analytical method are:

1. The dependence of the old-age constant C on ϵ is has been found, to the first power of ϵ, and agrees with numerical work.

2. An intriguing problem is the following: Knowing from numerical work that $C \approx 0.34$ in the limit $\epsilon \to 0$, express C by means of rational numbers and π etc. The partial solution of this problem in the form (6.95) and (6.96) gives an understanding that cannot be obtained by numerical work only.

6.3 The decay of a shockwave from a supersonic projectile

The solution (6.94) of the generalized Burgers' equation for an initial cylindrical N-wave can be used for calculating the decayed shockwave from a supersonic projectile. The parameters of the initial cylindrical N-wave depend on the shape and velocity of the projectile. The effects of nonlinearity accumulates during the propagation of the wave.

This is why nonlinearity may be neglected at short distances from the source. The nonlinear theory of the wave from a supersonic projectile is treated by Whitham (1950, 1952) for the case of no dissipation. In order to find the appropriate boundary condition of the nonlinear cylindrical wave we therefore begin with studying the wave from a supersonic projectile modelled by linear acoustic theory.

6.3.1 Linear theory of the wave from a supersonic projectile

Modelling of the sound source by a pulsating volume

A projectile as source of sound can be modelled by a pulsating volume where mass is created and annihilated in such a way that there is in the long run no change of net mass. This phenomenon can be taken into account by changing the mass conserving equation (2.1) in the following way:

$$\frac{\partial}{\partial t}\int_V \rho\, dV = -\int \rho v_i\, dS_i + \frac{d}{dt} m_S(t) \tag{6.97}$$

where we have assumed that, in a point $\mathbf{r} = \mathbf{r}_S$ within the volume V, mass is not conserved but varies with time according the function $m_S(t)$. The differential form of (6.97) is

$$\frac{\partial}{\partial t}[\rho - m_S(t)\delta(\mathbf{r}-\mathbf{r}_S)] + \frac{\partial}{\partial x_i}(\rho v_i) = 0 \tag{6.98}$$

where $\delta(\mathbf{r}-\mathbf{r}_S)$ is the three-dimensional Dirac delta function. Using (6.98) instead of (2.6) together with (2.41) we obtain instead of (2.55):

$$\frac{D}{Dt}[h - \gamma T_0(s-s_0)] - (\gamma-1)\frac{h}{\rho}[\rho\Delta\Phi + \dot{m}_S(t)\delta(\mathbf{r}-\mathbf{r}_S)] = 0\ . \tag{6.99}$$

The equation (2.56) then obtains an additional term:

$$\frac{\partial h}{\partial t} - \mathrm{grad}\Phi\cdot\mathrm{grad}\frac{\partial\Phi}{\partial t} - \frac{\gamma\kappa}{\rho_0 c_p}\Delta\frac{\partial\Phi}{\partial t}$$
$$-(\gamma-1)(c_pT_0 + \frac{\partial\Phi}{\partial t})\Delta\Phi - \frac{\gamma-1}{\rho}(c_pT_0 + \frac{\partial\Phi}{\partial t})\dot{m}_S(t)\delta(\mathbf{r}-\mathbf{r}_S) = 0\ . \tag{6.100}$$

Using (6.100) instead of (2.56) in the derivation of a generalized form of (2.60) we obtain after retaining only the linear lowest-order terms:

$$\frac{\partial^2 \Phi}{\partial t^2} - c_0^2 \Delta \Phi - \frac{c_0^2}{\rho} \dot{m}_S(t) \delta(\mathbf{r} - \mathbf{r}_S) = 0 \; . \tag{6.101}$$

The wave equation (6.101) describes the sound from a pulsating point mass in the point $\mathbf{r} = \mathbf{r}_S$. This case will be generalized to the case where the mass is extended along the x-axis and moves in the same direction. This means that the following exchange shall be made in (6.101):

$$m_S(t)\delta(x - x_S) \rightarrow \rho_0 a(x,t) \tag{6.102}$$

where ρ_0 is the mass density of the source (which is the same as the undisturbed mass density of the fluid) and is assumed to be constant. This means that $a(x,t)$ is the intersection area of a body which is rotationally symmetric around the x-axis and so slender that in practice the factor $\delta(y - y_S)\delta(z - z_S)$ with $y_S = z_S = 0$ can be used in the source term of the wave equation. The correctness of (6.102) is seen by the exchanges

$$\begin{aligned} \rho A &\rightarrow m_S \delta(x - x_S)\delta(y - y_S)\delta(z - z_S)a \\ &\rightarrow m_S \delta(x - x_S) \int_a \delta(y - y_S)\delta(z - z_S)\, dy\, dz \\ &= m_S \delta(x - x_S) \; . \end{aligned} \tag{6.103}$$

We now assume that the projectile has the constant velocity V and the intersection area $A(\xi)$ at the distance ξ behind the front. Then

$$a(x,t) = A(Vt - x) \; , \tag{6.104}$$

where the origin of the coordinate system is chosen so that the front of the projectile is at $x = 0$ at the time $t = 0$. The function $A(\xi)$ is different from zero for ξ values between 0 and L, where L is the length of the projectile (see Figure 6.1).

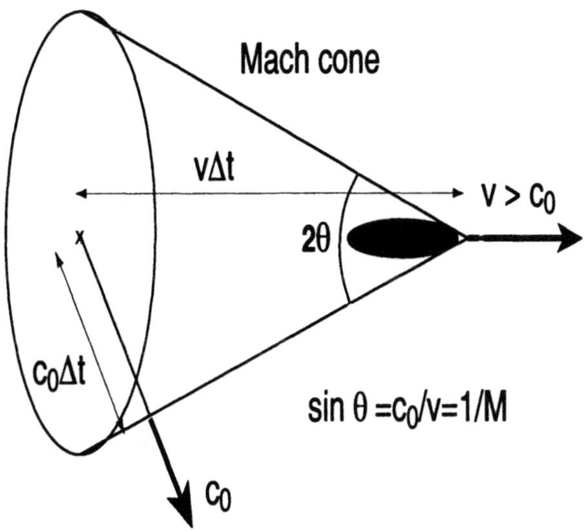

Figure 6.1: A Mach cone.

Sound field equation. Von Kármán's acoustical analogy.

From (6.101), (6.102 and (6.103) the equation of the sound field from a supersonic projectile in the linear region is obtained:

$$\frac{\partial^2 \Phi}{\partial t^2} - c_0^2 \Delta \Phi = c_0^2 \dot{A}(Vt - x)\, \delta(y)\, \delta(z) \tag{6.105}$$

The righthand side of (6.105) is invariant under the translation

$$t \to t + \Delta t\ , \quad x \to x + V\Delta t\ . \tag{6.106}$$

The invariance (6.106) is also valid for the derivatives on the lefthand side of (6.105). It must therefore be valid for Φ, so that Φ depends on x and t in the combination $Vt - x$. With the substitution

$$t_1 = t - \frac{x}{V} \tag{6.107}$$

the equation (6.105) becomes

$$\frac{\partial^2 \Phi}{\partial y^2} + \frac{\partial^2 \Phi}{\partial z^2} - (\frac{1}{c_0^2} - \frac{1}{V^2})\frac{\partial^2 \Phi}{\partial t_1^2} = -\frac{d}{dt_1} A(Vt_1)\, \delta(y)\, \delta(z)\ . \tag{6.108}$$

For $V > c_0$ the equation (6.108) is formally a wave equation for two-dimensional waves with wave velocity c^* defined as

$$c^* = \frac{V}{\sqrt{M^2 - 1}} , \tag{6.109}$$

where M is the Mach number (Ernst Mach (1838-1917), Austrian physicist and philosopher) for the projectile:

$$M = \frac{V}{c_0} . \tag{6.110}$$

The reformulation (6.108) of the linear supersonic wave problem is called von Kármán's (1947) acoustical analogy.

Green function method solution of the sound field equation

A solution of (6.108) is given by A.D. Pierce (1981) and some hints are given here. Equation (6.108) is treated as a special case of the corresponding wave equation with three space variables. This equation is solved by use of the Green function

$$G(\mathbf{r}, t; \mathbf{r}_0, t_0) = \frac{\delta(t - t_0 - \frac{|\mathbf{r}-\mathbf{r}_0|}{c_0})}{4\pi|\mathbf{r} - \mathbf{r}_0|} , \tag{6.111}$$

which is a solution to the equation

$$(\Delta - \frac{1}{c_0^2}\frac{\partial^2}{\partial t^2})G(\mathbf{r}, t; \mathbf{r}_0, t_0) = -\delta(\mathbf{r} - \mathbf{r}_S)\delta(t - t_0) \tag{6.112}$$

and vanishes at infinity.

Using (6.112) and (6.111) we find the solution, vanishing at infinity, of the equation

$$(\Delta - \frac{1}{c_0^2}\frac{\partial^2}{\partial t^2})\Phi(\mathbf{r}, t) = -4\pi s(\mathbf{r}, t) , \tag{6.113}$$

where the source function $s(\mathbf{r}, t)$ is different from zero in a limited region of space. The solution of (6.113) is then

$$\Phi(\mathbf{r}, t) = 4\pi \int\int\int\int G(\mathbf{r}, t; \mathbf{r}_0, t_0)s(\mathbf{r}_0, t_0)\, dx_0\, dy_0\, dz_0\, dt_0 \tag{6.114}$$

or

$$\Phi(\mathbf{r},t)=\int\int\int\frac{s(\mathbf{r}_0,t-\frac{|\mathbf{r}-\mathbf{r}_0|}{c_0})}{|\mathbf{r}-\mathbf{r}_0|}\,dx_0\,dy_0\,dz_0\ . \tag{6.115}$$

In order to use the solution formula (6.115) for solving (6.108) we introduce in (6.108) a ficticious coordinate x^*:

$$\frac{\partial^2\Phi}{\partial x^{*2}}+\frac{\partial^2\Phi}{\partial y^2}+\frac{\partial^2\Phi}{\partial z^2}-\frac{1}{c^{*2}}\frac{\partial^2\Phi}{\partial t_1^2}=-\frac{d}{dt_1}A(Vt_1)\,\delta(y)\,\delta(z) \tag{6.116}$$

Then we solve (6.116) in analogy with (6.113) using (6.115):

$$\Phi(x^*,y,z,t)=\frac{V}{4\pi}\int_{-\infty}^{\infty}\frac{A'(V(t_1-\frac{R}{c^*}))}{R}\,dx_S\ , \tag{6.117}$$

where A' means derivation with respect to the argument of A and R is given as

$$R=\sqrt{(x^*-x_S)^2+y^2+z^2}\ . \tag{6.118}$$

By changing the integration variable in (6.117) from x_S to $X=x_S-x^*$ we see that the righthand side of (6.117) does not depend on x^*. Thus $\Phi(x^*,y,z,t,)$ must be identical with our desired solution $\Phi(y,z,t)$ of (6.108). Changing the integration variable in (6.117) to ξ, given as

$$\xi=(t_1-\frac{R}{c^*})V\ , \tag{6.119}$$

and using the cylindrical symmetry with

$$r=\sqrt{y^2+z^2} \tag{6.120}$$

we find from (6.117):

$$\Phi(r,Vt-x)=\frac{V}{2\pi}\int_{-\infty}^{\xi_m}\frac{A'(\xi)\,d\xi}{\sqrt{(Vt-x-\xi)^2-(M^2-1)r^2}}\ . \tag{6.121}$$

The upper limit ξ_m of the integral in (6.121) is given as

$$\xi_m=Vt-x-(M^2-1)^{\frac{1}{2}}r=V(t-\frac{\mathbf{n}\cdot\mathbf{r}}{c_0})\ ,\qquad \mathbf{r}=x\mathbf{e}_x+r\mathbf{e}_r\ , \tag{6.122}$$

where

$$\mathbf{n}=\frac{1}{M}\mathbf{e}_x+\frac{(M^2-1)^{\frac{1}{2}}}{M}\mathbf{e}_r\ . \tag{6.123}$$

The angle between $\mathbf{n}$ and the projectile path, which is the x-axis, is $\arccos(\frac{1}{M})$.

The Mach cone. Whitham's F-function

Because $A'(\xi)$ is zero except for $0 < \xi < L$, the integral in (6.20) vanishes unless $\xi_m > 0$. This means that the effect of the projectile can be measured, as follows from (6.122), (6.123), in the domain

$$(M^2-1)^{\frac{1}{2}} r < Vt - x \ . \tag{6.124}$$

The domain given by (6.124) is a cone whose apex is situated on the x-axis at $x = Vt$. The cone is called the Mach cone (Figure 6.1). The opening angle of the Mach cone is $2\theta_M$, where

$$\theta_M = \arccos \frac{(M^2-1)^{\frac{1}{2}}}{M} \ . \tag{6.125}$$

The integral (6.121), which is the exact solution of the problem of calculating the sound field from a supersonic projectile in the linear region, can be approximated if the dimensions of the body are small compared with the distance from the body to the place where the sound is received. For this purpose we write the denominator in the integral in (6.121) using (6.122):

$$\begin{aligned}
&[(Vt-x-\xi)^2 - (M^2-1)r^2]^{\frac{1}{2}} \\
&= [(\xi_m + (M^2-1)^{\frac{1}{2}} r - \xi)^2 - (M^2-1)r^2]^{\frac{1}{2}} \\
&= (\xi_m - \xi)^{\frac{1}{2}} [2(M^2-1)^{\frac{1}{2}} r + \xi_m - \xi]^{\frac{1}{2}} \ .
\end{aligned} \tag{6.126}$$

The integral (6.121) is now approximated, using (6.126) and neglecting ξ and ξ_m in comparison with $2(M^2-1)^{\frac{1}{2}} r$:

$$\begin{aligned}
\Phi(r, Vt-x) &= \frac{V}{2\pi\sqrt{2}(M^2-1)^{\frac{1}{4}}} r^{-\frac{1}{2}} \int_{-\infty}^{\xi_m} \frac{A'(\xi)\, d\xi}{(\xi_m - \xi)^{\frac{1}{2}}} \\
&= \frac{V}{2\pi\sqrt{2}(M^2-1)^{\frac{1}{4}}} r^{-\frac{1}{2}} \int_0^{\infty} \frac{A'(\xi_m - \eta)\, d\eta}{\eta^{\frac{1}{2}}} \ .
\end{aligned} \tag{6.127}$$

The approximation made in (6.127) is founded on the inequality $0 < \xi < L$, and the assumptions that $L \ll (M^2+1)^{\frac{1}{2}} r$ and that ξ_m is of the same order of magnitude as L. The integral in (6.127) depends only on the geometry of the sound source. It is interesting to note in (6.127), using (6.122), that the velocity potential Φ decays as $r^{-\frac{1}{2}}$ for constant $t - \mathbf{n} \cdot \mathbf{r}/c_0$. The connection between Φ and the pressure p is, in the lowest order, given by (2.47):

$$p = \rho_0 \frac{\partial \Phi}{\partial t} , \tag{6.128}$$

where p means the deviation from the equilibrium pressure. Thus from (6.127) and (6.122) we obtain

$$p(r, Vt - x) = \frac{V^2 \rho_0}{\sqrt{2}(M^2 - 1)^{\frac{1}{4}} r^{\frac{1}{2}}} F_W(Vt - x - (M^2 - 1)^{\frac{1}{2}} r) , \tag{6.129}$$

where Whitham's F-function F_W is defined as (Whitham 1974; Pierce 1981)

$$F_W(\xi) = \frac{1}{2\pi} \frac{d^2}{d\xi^2} \int_0^\infty \frac{A(\xi - \eta)\, d\eta}{\eta^{\frac{1}{2}}} . \tag{6.130}$$

We note that $F_W(\xi) = 0$ for $\xi < 0$. We assume that $A(\xi)$ approaches zero faster than $\xi^{\frac{3}{2}}$ for $\xi \to 0$. Then $F_W(\xi)$ is finite for $\xi = 0$. The result (6.129) is derived by Pierce (1981), formula (11-10.9).

6.3.2 Nonlinear theory of the wave from a supersonic projectile

Generalized Burgers' equation of the sound wave from a supersonic projectile

After having travelled a certain distance $r > r_0$, the wave from the projectile calculated in subsection 6.3.1 must be described by nonlinear theory. For this description we start with Kuznetsov's equation (2.60). Assuming cylindrical symmetry we search for a solution of (2.60) depending on r and $Vt - x$ like the solution in (6.121). In (2.60) we make the substitution (6.120) and the substitutions (Enflo 1985[b,c])

$$X = c_0(t - \frac{V}{c_0^2} x) \tag{6.131}$$

$$T = \frac{1}{c_0}(x - Vt) . \tag{6.132}$$

Like the wave calculated in (6.127) the nonlinear wave depends on r and $Vt - x$ only. Therefore we require

$$\frac{\partial \Phi(r, X, T)}{\partial X} = 0 \ . \tag{6.133}$$

Use of (6.131)-(6.133) in (2.60) yields

$$\begin{aligned} &\frac{\partial^2 \Phi}{\partial r^2} + \frac{1}{r}\frac{\partial \Phi}{\partial r} - \frac{M^2 - 1}{c_0^2}\frac{\partial^2 \Phi}{\partial T^2} \\ &= \frac{2V}{c_0^5}(1 - \frac{\gamma - 1}{2}M^2)\frac{\partial \Phi}{\partial T}\frac{\partial^2 \Phi}{\partial T^2} + \frac{V}{c_0^5}\frac{b}{\rho_0}\frac{\partial^3 \Phi}{\partial T^3} \\ &\quad + \frac{V}{c_0^3}\frac{\partial}{\partial T}(\frac{\partial \Phi}{\partial r})^2 + \frac{b}{\rho_0}\frac{V}{c_0^3}\frac{\partial}{\partial T}(\frac{\partial^2 \Phi}{\partial r^2} + \frac{1}{r}\frac{\partial \Phi}{\partial r}) \ . \end{aligned} \tag{6.134}$$

New variables are now introduced in (6.134):

$$\tau = -T - \frac{r - r_0}{c_0}(M^2 - 1)^{\frac{1}{2}} = \frac{1}{c_0}[Vt - x - (M^2 - 1)^{\frac{1}{2}}(r - r_0)] \tag{6.135}$$

$$r_1 = \nu r \ , \tag{6.136}$$

where ν is a small dimensionless parameter. We will see that the final result is insensitive to the choice of r_0, the distance from the flight path for which the accumulated effects of nonlinearity become important. Inserting (6.135) and (6.136) into (6.134) and expanding to second order in the small quantities ν, Φ and b we obtain a generalized Burgers' equation:

$$\begin{aligned} &\frac{\partial v}{\partial r} + \frac{1}{2}\frac{v}{r} - \frac{M^3}{c_0^2(M^2 - 1)}\frac{\gamma + 1}{2}v\frac{\partial v}{\partial \tau} \\ &= \frac{M^3 b}{2c_0^3 \rho_0 \sqrt{M^2 - 1}}\frac{\partial^2 v}{\partial \tau^2} \end{aligned} \tag{6.137}$$

where

$$v = v_r = -\frac{\partial \Phi}{\partial r} = \frac{\sqrt{M^2 - 1}}{c_0}\frac{\partial \Phi}{\partial \tau} - \nu\frac{\partial \Phi}{\partial r_1} \ . \tag{6.138}$$

The equation (6.137) is a generalization of (2.85) with $n = 1$. Because M is greater than 1 in (6.137) we cannot obtain (2.85) by a limit transition in (6.137). We note that (6.137) is suitable for a boundary value problem, while the equation (6.6), for which we have obtained the asymptotic wave in section 5.2, has its origin in an initial value problem. From a mathematical point of view the two problems are identical.

The generalized Burgers' equation (6.137) will now be studied with an appropriate boundary condition, derived from (6.129). To that end we use (6.128), (6.132), (6.135) and the dependence of p on r and $Vt-x$ given in (6.129) and find

$$p = \rho_0 \left(\frac{\partial \Phi}{\partial t}\right)_{x,y,z} = -\rho_0 \frac{V}{c_0}\left(\frac{\partial \Phi}{\partial T}\right)_r = \rho_0 \frac{V}{c_0}\left(\frac{\partial \Phi}{\partial \tau}\right)_r . \tag{6.139}$$

Comparison of (6.139) and (6.138) gives in the lowest order of ν:

$$v = \frac{p}{\rho_0 c_0}\frac{\sqrt{M^2-1}}{M} . \tag{6.140}$$

Using (6.129), (6.135) and (6.140) the fluid velocity at the boundary $r = r_0$ is found:

$$v(r = r_0, \tau) = \frac{V(M^2-1)^{\frac{1}{4}}}{(2r_0)^{\frac{1}{2}}} F_W\left(c_0\left(\tau - \frac{r_0}{c_0}\sqrt{M^2-1}\right)\right) . \tag{6.141}$$

An example of the function F_W, generalized by a projectile whose surface is a rotational surface of a parabola, is given in Figure 6.3. From this figure it is seen that F_W is zero outside the length of which slightly exceeds L, and has a zero with negative derivative in the middle of this interval. In order that the position of this zero obtains zero coordinate value, we make the following coordinate translation:

$$\theta = \tau - \frac{r_0}{c_0}\sqrt{M^2-1} - \frac{L}{2c_0} . \tag{6.142}$$

The term without derivative in (6.137) is eliminated by the transformations

$$U = v\left(\frac{r}{r_0}\right)^{\frac{1}{2}} \tag{6.143}$$

$$\zeta = 2(rr_0)^{\frac{1}{2}} . \tag{6.144}$$

Substituting (6.142), (6.143), (6.144) into (6.137) we obtain

$$\frac{\partial U}{\partial \zeta} - \frac{M^3}{M^2-1}\frac{\gamma+1}{2c_0^2}U\frac{\partial U}{\partial \theta} = \frac{M^2}{\sqrt{M^2-1}}\frac{b\zeta}{4c_0^3\rho_0 r_0}\frac{\partial^2 U}{\partial \theta^2} . \tag{6.145}$$

The equation (6.145) is a generalization to the case $M > 1$ of equation (1.6) in chapter 3 of Rudenko and Soluyan (1977).

Transformation of a Whitham function boundary value problem to an N-wave boundary value problem

The N-wave boundary value problem of (6.145) is, from a mathematical point, the same problem as the N-wave initial value problem of (6.20) with (6.23). This problem has been solved in section 5.2 and the solution found there can be applied to (6.145), if we apply an N-wave boundary condition to (6.145). The boundary wave (6.141) is in general not an N-wave. However, it will develop into an N-wave, as will be shown.

For the description of the evolution of the wave (6.141) into an N-wave we use the inviscid equation found by neglecting the righthand side of (6.145):

$$\frac{\partial U}{\partial \zeta} - \Gamma U\frac{\partial U}{\partial \theta} = 0 , \qquad \Gamma = \frac{M^3}{M^2-1}\frac{\gamma+1}{2c_0^2} . \tag{6.146}$$

The boundary condition (6.141) gives with (6.142), (6.143) and (6.144):

$$U(\zeta - 2r_0 = 0 , \theta) = G(\theta) , \tag{6.147}$$

where

$$G(\theta) = \frac{V(M^2-1)^{\frac{1}{4}}}{(2r_0)^{\frac{1}{2}}} F_W(c_0(\theta + \frac{L}{2c_0})) . \tag{6.148}$$

The solution of (6.146) with the boundary condition (6.147) is constructed in analogy with (4.33), (4.34):

$$U = G(\Psi) , \tag{6.149}$$

where Ψ is implicitly given as a function of τ and ζ through the relation

$$\Psi = \theta + \Gamma G(\Psi)(\zeta - 2r_0) \tag{6.150}$$

For the following principial discussion we choose a somewhat idealized form of $G(\theta)$, shown in Figure 6.2.

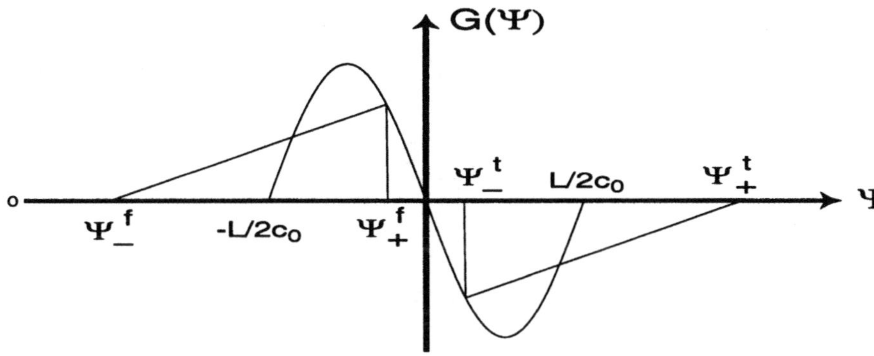

Figure 6.2: Construction of positions for front shock and tail shock

The function $G(\theta)$ fulfils the zero area condition

$$\int_{-\infty}^{\infty} G(\Psi)\, d\Psi = 0 \tag{6.151}$$

and develops into an N-wave with a positive front shock discontinuity, a linear decrease and a positive tail shock discontinuity. Because of the antisymmetry of the wave around $\Psi = 0$ it is sufficient to study the front shock. The position θ_{fsh} of the front shock is given in analogy with (3.39):

$$\theta_{fsh} = \Psi_-^f - \Gamma G(\Psi_-^f)(\zeta - 2r_0) = \Psi_+^f - \Gamma G(\Psi_+^f)(\zeta - 2r_0)\ . \tag{6.152}$$

From Figure 6.2 it is obvious that $G(\Psi_-^f) = 0$ and thus according to (6.152) $\Psi_-^f = \theta_{fsh}$. The rule of equal areas (3.53) applied to the front shock becomes

$$\int_{\Psi_-^f}^{\Psi_+^f} G(\Psi)\, d\Psi = \frac{1}{2}(\Psi_+^f - \Psi_-^f)(G(\Psi_+^f) + G(\Psi_-^f))\ . \tag{6.153}$$

Because we have from (6.152)

$$\Psi_+^f - \Psi_-^f = \Gamma(\zeta - 2r_0)(G(\Psi_+^f) - G(\Psi_-^f)) \tag{6.154}$$

and using $G(\Psi_-^f) = 0$ we can write (6.153):

$$\int_{-\infty}^{\Psi_+^f} G(\Psi)\, d\Psi \equiv I_+^f = \frac{1}{2}\Gamma(\zeta - 2r_0)G^2(\Psi_+^f)\ . \tag{6.155}$$

For the discontinuity $U_{fsh} = G(\Psi_+^f)$ of the front shock the equation (6.155) gives

$$U_{fsh} = \left[\frac{2I_+^f}{\Gamma(\zeta - 2r_0)}\right]^{\frac{1}{2}}\ . \tag{6.156}$$

For ζ sufficiently great the solution (6.149), (6.150), with due consideration of the discontinuities, approaches an N-wave.

Example of N-wave generation

In order to study the generation of an N-wave from the wave developing from (6.147), (6.148) we choose an example of the function $G(\Psi)$:

$$\left.\begin{aligned} &G(\Psi) = -U_0 \sin \Omega\Psi\,, && -\pi < \Omega\Psi < \pi\,,\ \ \Omega = 2\pi c_0/L = 2\pi V/(ML) \\ &G(\Psi) = 0\,, && -\infty < \Omega\Psi < -\pi\,,\ \ \pi < \Omega\Psi < \infty\,. \end{aligned}\right\} \tag{6.157}$$

The rule of equal areas (6.153) becomes with (6.157)

$$\frac{1}{2}(\Psi_+^f - \Psi_-^f)\big(-U_0 \sin(\Omega\Psi_+^f)\big) = \int_{-\pi/\Omega}^{\Psi_+^f} [-U_0 \sin(\Omega\Psi)]\, d\Psi \tag{6.158}$$

or

$$\Psi_-^f \sin(-\Omega\Psi_+^f) = \Psi_+^f \sin(-\Omega\Psi_+^f) + \frac{2}{\Omega}(-1 - \cos(\Omega\Psi_+^f))\ . \tag{6.159}$$

From (6.152) and (6.157) we obtain

$$\Psi_+^f = \Psi_-^f - \Gamma U_0 \sin(\Omega\Psi_+^f)(\zeta - 2r_0) \tag{6.160}$$

We will solve Ψ_+^f and Ψ_-^f from the system of equations (6.159) and (6.160). To this end we assume that $\Omega\Psi_+^f$ is so small that we can use the approximations

$$\sin \Omega\Psi_+^f \approx \Omega\Psi_+^f\,, \qquad \cos \Omega\Psi_+^f \approx 1 - \frac{1}{2}(\Omega\Psi_+^f)^2 \tag{6.161}$$

In the lowest order the solution of (6.159), (6.160) with the approximations (6.161) is

$$\Psi_+^f = -\frac{2}{\Omega^{\frac{3}{2}}\Gamma^{\frac{1}{2}}U_0^{\frac{1}{2}}(\zeta - 2r_0)^{\frac{1}{2}}} \tag{6.162}$$

$$\Psi_-^f = \theta_{fsh} = -\frac{2\Gamma^{\frac{1}{2}}U_0^{\frac{1}{2}}(\zeta - 2r_0)^{\frac{1}{2}}}{\Omega^{\frac{1}{2}}} . \tag{6.163}$$

The equation of the N-wave is then found from (6.149), (6.150):

$$U = -U_0 \sin\Omega\Psi = \frac{\Psi - \theta}{\Gamma(\zeta - 2r_0)} . \tag{6.164}$$

Because $\sin\Omega\Psi \approx \Omega\Psi$ we have from (6.164)

$$\Psi \approx -\frac{U}{U_0\Omega} \tag{6.165}$$

and thus from (6.164) in the lowest order of $(\zeta - 2r_0)^{-1}$

$$U = -\frac{\theta}{\Gamma(\zeta - 2r_0)} , \tag{6.166}$$

where θ lies in the interval

$$-\frac{2\Gamma^{\frac{1}{2}}U_0^{\frac{1}{2}}(\zeta - 2r_0)^{\frac{1}{2}}}{\Omega^{\frac{1}{2}}} < \theta < \frac{2\Gamma^{\frac{1}{2}}U_0^{\frac{1}{2}}(\zeta - 2r_0)^{\frac{1}{2}}}{\Omega^{\frac{1}{2}}} . \tag{6.167}$$

For θ outside the interval (6.167) we have

$$U = 0 . \tag{6.168}$$

The height of the front shock is found from (6.166) and (6.167):

$$U_{fsh} = \frac{2U_0^{\frac{1}{2}}}{\Omega^{\frac{1}{2}}\Gamma^{\frac{1}{2}}(\zeta - 2r_0)^{\frac{1}{2}}} . \tag{6.169}$$

The result (6.169) is the same as (6.156) with (6.157).

Determination of the distance at which the wave is an N-wave

The derivation of the N-wave solution (6.166)-(6.168) of (6.146) depends on the assumption that $\Omega\Psi_+^f$ is sufficiently small for the approximations (6.161) to be valid. This is the case for sufficiently great ζ values. If we want to use (6.161) for $r > R_0$, i.e. $\zeta > 2R_0^{\frac{1}{2}} r_0^{\frac{1}{2}}$, then we must require

$$\frac{1}{6}\Omega^2\Psi_+^{f2} \ll 1 \tag{6.170}$$

in order that the next term in the expansion of the sine and cosine functions be negligible. Insertion of (6.162) into (6.170) and use of the expressions for Γ and Ω in (6.146) and (6.157) respectively give

$$\frac{(M^2-1)Lc_0}{6\beta M^3 \pi U_0 \sqrt{R_0 r_0}} \ll 1\,, \qquad \beta = \frac{\gamma+1}{2}\,. \tag{6.171}$$

An estimation of U_0 is found from (6.157), which tells us that U_0 is the maximum of $G(\theta)$. Because, according to (6.141), (6.142), (6.143) and (6.148),

$$U_0 = v_{max}(r = r_0, \tau)\ , \tag{6.172}$$

we find the result

$$U_0 = \frac{V(M^2-1)^{\frac{1}{4}}}{(2r_0)^{\frac{1}{2}}} F_W^{max}(c_0(\tau - \frac{r_0}{c_0}\sqrt{M^2-1}))\ . \tag{6.173}$$

In order to know how well the inequality (6.171) is satisfied we must insert (6.173) into (6.171). Thus we must have some realistic estimation of the maximum F_W^{max} of the Whitham function F_W defined in (6.130). We assume that the area function $A(\xi)$ in (6.130) describes a projectile with a circular cross section and put (Pierce 1981)

$$\begin{aligned} &A(\xi) = 4\pi R_{max}^2 \frac{\xi}{L}(1-\frac{\xi}{L})\,, && 0 \le \xi \le L \\ &A(\xi) = 0\ , \qquad \xi < 0\ , && \xi > L\ . \end{aligned} \tag{6.174}$$

Inserting (6.174) into (6.130) we obtain

$$\frac{L^{3/2}}{R_{max}^2} F_W(\frac{\xi}{L}) = 2\frac{d^2}{d(\xi/L)^2}\int_{\xi/L=0}^{\xi/L=1} [\frac{\xi-\eta}{L}(1-\frac{\xi-\eta}{L})]/(\frac{\eta}{L})^{1/2}\, d(\frac{\eta}{L})\ . \tag{6.175}$$

The dimensionless function $F_W(\xi/L)L^{3/2}/R_{max}^2$, evaluated by (6.175), is shown in Figure 6.3.

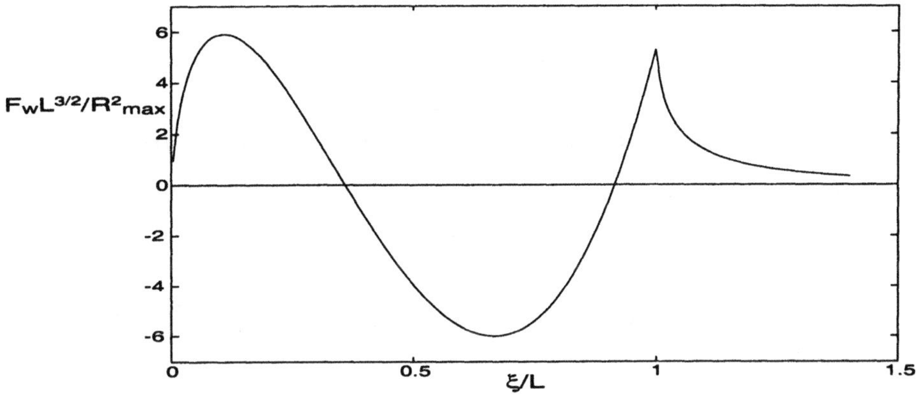

Figure 6.3: The function $F_W(\xi/L)L^{3/2}/R_{max}^2$ evaluated according to equation (6.130) with (6.174).

From Figure 6.3 we see that

$$F_W^{max}\frac{L^{3/2}}{R_{max}^2} \approx 6 \,, \tag{6.176}$$

where R_{max} is the maximal radius of the circular intersection area of the projectile. Substituting (6.173) and (6.176) in (6.171) we obtain

$$\frac{(M^2-1)^{\frac{3}{4}}}{\pi\beta M^4}\frac{\sqrt{2}}{36}\frac{L^{\frac{5}{2}}}{R_{max}^2 R_0^{\frac{1}{2}}} \ll 1 \,. \tag{6.177}$$

The maximum of $\frac{(M^2-1)^{\frac{3}{4}}}{M^4}$ is 0.2663 for $M^2 = 1.6$. With $\beta = 1.2$ for air we write (6.177):

$$0.00277\Big(\frac{L}{R_{max}}\Big)^2\Big(\frac{L}{R_0}\Big)^{\frac{1}{2}} \ll 1 \,. \tag{6.178}$$

A reasonable value of L/R_{max} is 10, which means that (6.178) can be written

$$\frac{1}{3.6}\Big(\frac{L}{R_0}\Big)^{\frac{1}{2}} \ll 1 \,. \tag{6.179}$$

If an error at the five percent level is accepted, Eq. (6.179) means that the ratio L/R_0 becomes

$$\frac{L}{R_0} = 3.6^2 \cdot \frac{1}{400} \tag{6.180}$$

so that for $L = 10$ m we find $R_0 \approx 300$ m.

The approximate N-wave (6.166)-(6.168) will now provide the boundary condition for $r = R_0$ (i.e. $\zeta = 2R_0^{\frac{1}{2}} r_0^{\frac{1}{2}}$) of the solution of the generalized Burgers' equation (6.145).

A dimensionless wave equation, boundary conditions and solution

In order to obtain a dimensionless equation we make a change of variables in (6.137) similar to (6.2)-(6.5). The equation (6.2) corresponds to

$$W = \left(\frac{r}{R_0}\right)^{\frac{1}{2}} \cdot \frac{v(r,\tau)}{v_0} = \left(\frac{r_0}{R_0}\right)^{\frac{1}{2}} \cdot \frac{U}{v_0} , \tag{6.181}$$

where v_0 is a velocity, which has to be chosen appropriately, as we will see below. Instead of (6.3) we have (cf. (6.142))

$$s = \frac{c_0}{L}\left(\tau - \frac{r_0}{c_0}\sqrt{M^2-1} - \frac{L}{2c_0}\right) = \frac{c_0}{L}\theta \tag{6.182}$$

and instead of (6.4) and (6.5) we introduce the dimensionless parameter η_0 and the dimensionless variable η as:

$$\eta_0 = 2\beta \frac{M^3}{M^2-1} \frac{v_0 R_0}{c_0 L} \tag{6.183}$$

$$\begin{aligned} \eta &= 1 + 2\beta \frac{M^3}{M^2-1} \frac{v_0 R_0^{\frac{1}{2}}}{c_0 L} \left(r^{\frac{1}{2}} - R_0^{\frac{1}{2}}\right) \\ &= 1 + 2\beta \frac{M^3}{M^2-1} \frac{v_0 R_0^{\frac{1}{2}}}{c_0 L} \left(\frac{\zeta}{2r_0^{\frac{1}{2}}} - R_0^{\frac{1}{2}}\right) . \end{aligned} \tag{6.184}$$

Insertion of (6.181)-(6.184) into (6.137) gives

$$\frac{\partial W}{\partial \eta} - W\frac{\partial W}{\partial s} = \epsilon \, \frac{\eta - 1 + \eta_0}{2} \frac{\partial^2 W}{\partial s^2} , \tag{6.185}$$

where the small dimensionless parameter ϵ is given as

$$\epsilon = \frac{\sqrt{M^2-1}\,b}{\beta\eta_0\rho_0 v_0 L} \tag{6.186}$$

The N-wave (6.166)-(6.168) will now give the boundary condition of (6.185) for $\eta = 1$ which means $\zeta = 2r_0^{\frac{1}{2}}R_0^{\frac{1}{2}}$. The condition for the boundary wave to be an approximate N-wave is (6.177), which is a condition for R_0 if the parameters β, M, L, R_{max} are given. The velocity v_0 is at our disposal for choosing the dimensionless front shock amplitude W_{fsh} to be unity at $\eta = 1$. Using (6.181), (6.184) and (6.173) in (6.169) we obtain

$$W = \left(\frac{r_0}{R_0}\right)^{\frac{1}{2}}\frac{U_{fsh}}{v_0} = \frac{2^{\frac{1}{4}}M^{\frac{1}{2}}(M^2-1)^{\frac{1}{8}}(F_W^{max})^{\frac{1}{2}}c_0^{\frac{1}{2}}}{\pi^{\frac{1}{2}}v_0^{\frac{1}{2}}R_0^{\frac{1}{4}}[\eta-1+\eta_0-\eta_0(\frac{r_0}{R_0})^{\frac{1}{2}}]^{\frac{1}{2}}} \,. \tag{6.187}$$

The condition

$$W_{fsh}(\eta = 1) = 1 \tag{6.188}$$

gives together with (6.183)

$$v_0 = \frac{(M^2-1)^{\frac{5}{8}}(F_W^{max})^{\frac{1}{2}}c_0 L^{\frac{1}{2}}}{2^{\frac{1}{4}}\pi^{\frac{1}{2}}\beta^{\frac{1}{2}}MR_0^{\frac{3}{4}}[1-(\frac{r_0}{R_0})^{\frac{1}{2}}]^{\frac{1}{2}}} \,. \tag{6.189}$$

From (6.189) and (6.183) we then obtain

$$\eta_0 = \frac{2^{\frac{3}{4}}\beta^{\frac{1}{2}}M^2(F_W^{max})^{\frac{1}{2}}R_0^{\frac{1}{4}}}{\pi^{\frac{1}{2}}(M^2-1)^{\frac{3}{8}}L^{\frac{1}{2}}[1-(\frac{r_0}{R_0})^{\frac{1}{2}}]^{\frac{1}{2}}} \,. \tag{6.190}$$

We remember that the condition (6.177) is of fundamental importance for the N-wave (6.166)-(6.168) to be generated and useful as a boundary condition of (6.185) for $\eta = 1$. We investigate how well (6.177) is fulfilled for the choice $\eta_0 = 1$. Using (6.190), (6.176) and (6.177) gives

$$\frac{2}{3\pi^2} \ll 1 \,, \tag{6.191}$$

which means an accuracy at the same level as (6.179), (6.180). Putting $\eta_0 = 1$ in (6.185) we can use the initial condition (6.26), (6.27) and the outer solution (6.28), (6.29) of the dimensionless generalized Burgers' equation (6.20) with (6.23). The boundary condition of (6.185) with $\eta_0 = 1$ thus is

$$W(s, \eta = 1) = -s , \qquad |s| < 1 \tag{6.192}$$

$$W(s, \eta = 1) = 0 , \qquad |s| > 1 \tag{6.193}$$

and the outer solution is

$$W(s, \eta = 1) = -\frac{s}{\eta} + O(\epsilon^n) \qquad |s| < \eta^{\frac{1}{2}} \tag{6.194}$$

$$W(s, \eta = 1) = 0 , \qquad |s| > \eta^{\frac{1}{2}} . \tag{6.195}$$

The inner solution analogous to (6.30)-(6.32) is

$$W^*(s^*, \eta) = -\frac{1}{2}\eta^{-\frac{1}{2}}\Big(1 - \tanh\frac{s^* - A(\eta)}{2\eta^{\frac{3}{2}}}\Big) + O(\epsilon) , \tag{6.196}$$

where

$$s^* = \frac{s - \eta^{\frac{1}{2}}}{\epsilon} \tag{6.197}$$

and

$$A(\eta) = -\frac{1}{2}\eta(\eta - 1) . \tag{6.198}$$

Putting $\eta_0 = 1$ in (6.190) we obtain an expression for R_0 depending only on given data of the air and the projectile. Using this expression in (6.189) we obtain from (6.186):

$$\epsilon = \frac{(M^2 - 1)\pi^2 b}{4\beta^2 M^5 \rho_0 c_0 (F_W^{max})^2} . \tag{6.199}$$

The expression (6.199) is valid for $G(\Psi)$ chosen according to (6.157) with U_0 given by (6.173). A still more general expression for ϵ is found using the formula, obtained from (6.148) and (6.155):

$$I_f^+(2r_0)^{\frac{1}{2}} = V(M^2 - 1)^{\frac{1}{2}} \int_{-\infty}^{\Psi_+^f} F_W(\theta + \frac{L}{2c_0})\, d\theta \tag{6.200}$$

and the formula, obtained from (6.155), (6.157) and (6.173):

$$I_f^+(2r_0)^{\frac{1}{2}} = \frac{M(M^2-1)^{\frac{1}{2}}L}{\pi} F_W^{max} . \tag{6.201}$$

Use of (6.200) and (6.201) in (6.199) gives a new expression for ϵ:

$$\epsilon = \frac{(M^2-1)^{\frac{1}{2}}L^2 b}{4\beta^2 M^3 \rho_0 c_0 V^2 (I_W^+)^2} , \tag{6.202}$$

where I_W^+ is defined as the integral

$$I_W^+ = \int_{-\infty}^{\Psi_+^f} F_W(\theta + \frac{L}{2c_0})\, d\theta , \qquad \Psi_+^f \to 0 \tag{6.203}$$

In practise, the integral in (6.203) means the area under the positive part of the Whitham function F_W.

The asymptotic solution of (6.185) with $\eta_0 = 1$ for $\eta \gg \frac{1}{\epsilon}$ can now be written in analogy with (6.94):

$$W(\eta, s) \approx -\epsilon^{-2} C \frac{s}{\eta^3} \exp(-\frac{s^2}{\epsilon\eta^2}) , \tag{6.204}$$

where C is the old-age constant discussed in connection with (6.95), (6.96).

Asymptotic solution expressed in physical quantities

It remains to express W, s, η in (6.204) by means of given physical quantities. From (6.190) with $\eta_0 = 1$, (6.189), (6.200), (6.201) and (6.203) we obtain (with $r_0 \to 0$)

$$R_0^{\frac{1}{2}} v_0 = \frac{(M^2-1)^{\frac{1}{4}} c_0 V I_W^+}{2^{\frac{1}{2}} L} . \tag{6.205}$$

Using (6.205) we find from (6.181) and (6.184), putting $\eta_0 = 1$ and $r_0 \to 0$:

$$W = \frac{2^{\frac{1}{2}} L}{(M^2-1)^{\frac{1}{4}} c_0 V I_W^+} r^{\frac{1}{2}} v(r,\tau) \tag{6.206}$$

$$\eta = \frac{2^{\frac{1}{2}} \beta M^3 V I_W^+}{(M^2-1)^{\frac{3}{4}} L^2} r^{\frac{1}{2}} . \tag{6.207}$$

For s we use in (6.204) the expression given in (6.182). Thus, with (6.202), (6.203), (6.206) and (6.207), the equation (6.204) gives an asymptotic form of $v(r, \tau)$ for $r \to \infty$.

Discussion

Now a summary is made of the problem of finding the dependence on the distance from the flight path of the asymptotic wave from a supersonic projectile. We compare three cases:

1. Linear theory without dissipation.

2. Nonlinear theory without dissipation.

3. Nonlinear theory with dissipation.

The r dependence of the asymptotic wave in case 1 is given in (6.129), which shows the asymptotic dependence $r^{-1/2}$. We expect that taking nonlinearity into account gives a faster decay. Such a decay is found by Whitham (1974), where the dependence $r^{-3/4}$ in case 2 is given. From the final formulae (6.204)-(6.207) the asymptotic r dependence of $v(r, \tau)$ in case 3 can be found. The result is r^{-1}. Of course it is expected that taking both nonlinearity and dissipation into account gives the fastest decay of the wave. A supersonic aircraft produces a sonic boom. The above treatment is relevant for the sonic boom problem; however there are many aspects of this problem besides the propagation of a cylindrical wave in unlimited space. Some of these aspects are treated by Pierce (1993). The sonic boom problem under the influence of a turbulent layer is studied by Rudenko and Enflo (2000).

6.4 Periodic cylindrical and spherical waves

As example of the evolution of a nonlinear periodic wave governed by a Burgers-like equation having no appropriate exact solution we choose the boundary value problem for cylindrical and spherical sinusoidal waves. The appropriate wave equations are given in (6.18) for spherical waves and in (6.15) for cylindrical waves. In both these cases the wave equation contains two parameters. In addition to the inverse acoustical Reynolds number the problem contains one more parameter because of the existence of a radius of the cylindrical or spherical source. This new parameter corresponds to the new parameter T_0 in the cylindrical initial wave problem of (6.6).

6.4.1 Spherical periodic waves

Approximate Fay solution

We begin with the spherical wave problem, since it is more documented in the literature (Shooter, Muir and Blackstock 1974), and study the equation given in (6.18), with $\epsilon^{(4)} = \epsilon$,

$$\frac{\partial V}{\partial R} - V\frac{\partial V}{\partial \theta} = \epsilon \exp\left(\frac{R}{R_0}\right)\frac{\partial^2 V}{\partial \theta^2} \tag{6.208}$$

with the sinusoidal boundary condition for $r = r_0$, i.e. $R = 0$:

$$V(0, \theta) = \sin\theta \; . \tag{6.209}$$

The problem of finding an approximate Fourier expansion of the solution of (6.208) can now be solved using the solution for the corresponding plane wave problem found in chapter 5. For $R < 1$ and $\epsilon \ll 1$ we thus have the Bessel-Fubini solution (cf. (4.46))

$$V = 2\sum_{n=1}^{\infty} \frac{J_n(nR)}{nR} \sin n\theta \; . \tag{6.210}$$

For $R \gg 1$ and $\epsilon \ll 1$ we can use the procedure described in (5.151)-(5.160), leading to the Khokhlov-Soluyan solution (5.160). We only have to make the exchange $\epsilon \to \epsilon\exp(\frac{R}{R_0})$. The resulting approximate Khokhlov-Soluyan solution of (6.208) with (6.209) thus is

$$V = \frac{1}{R+1}\Big(-\theta + \pi \tanh \frac{\pi\theta\exp(-\frac{R}{R_0})}{2\epsilon(R+1)}\Big)(1 + O(\epsilon)) \; . \tag{6.211}$$

In order to make the comparison with results of other authors easier we make the change of variable

$$z = R/R_0 \tag{6.212}$$

and introduce the new constant a as

$$a = \epsilon R_0 \,, \qquad a = O(1) \,. \tag{6.213}$$

Thus for the spherical Burgers' equation

$$\frac{\partial V}{\partial z} - a\epsilon^{-1} V \frac{\partial V}{\partial \theta} = a e^z \frac{\partial^2 V}{\partial \theta^2} \tag{6.214}$$

with the boundary condition

$$V(z = 0, \theta) = \sin\theta \tag{6.215}$$

we have the approximate Khokhlov-Soluyan solution

$$V = \frac{\epsilon}{\epsilon + az}\Big(-\theta + \pi \tanh \frac{\pi e^z \theta}{2(\epsilon + az)}\Big)(1 + O(\epsilon)) \tag{6.216}$$

and the approximate Fay solution (cf. (5.85))

$$V = 2\epsilon e^z \sum_{n=1}^{\infty} \frac{\sin n\theta}{\sinh[\epsilon\, e^z n(1 + a\epsilon^{-1}z)]}(1 + O(\epsilon)) \,. \tag{6.217}$$

Insertion of (6.216) into (6.214) shows that the requirement for (6.216) to be a good approximation is

$$\frac{\epsilon + az}{a} \ll 1 \,. \tag{6.218}$$

The inequality (6.218) gives the z region for which (6.216) is useful. Since a is $O(1)$ we choose

$$z = \delta\zeta \,, \qquad \frac{\epsilon}{\delta} \ll 1 \,, \qquad \delta \ll 1 \,. \tag{6.219}$$

The series (6.217) can be written using (6.219):

$$V = 4\epsilon \sum_{n=1}^{\infty} \sin n\theta \sum_{\nu=0}^{\infty} \exp[-(1 + 2\nu)n(\epsilon + \delta a\zeta)](1 + O(\delta)) \,. \tag{6.220}$$

Construction of a series solution from the asymptotic solution

By using the series (6.220) we will now determine the amplitude constant of the solution of the asymptotic equation (cf. (6.214))

$$\frac{\partial V}{\partial z} = ae^{z}\frac{\partial^2 V}{\partial \theta^2} , \tag{6.221}$$

where z is great enough that the nonlinear nonlinear term in (6.214) can be neglected. The solution of (6.221) shall develop from (6.216) and has the form

$$V = \frac{\epsilon}{a} C \exp(-ae^{z}) \sin\theta . \tag{6.222}$$

It is easily seen that (6.222) satisfies (6.221) identically. From (6.222) we obtain the first term in the series solution of (6.214) (Enflo 1996):

$$V = \frac{\epsilon}{a} C\exp(-ae^{z})[f_1(\theta) + \exp(-ae^{z})f_2(z,\theta) + \exp(-2ae^{z})f_3(z,\theta) + \ldots] . \tag{6.223}$$

By inserting (6.223) into (6.214) we find that a solution of (6.214) is obtained if the following differential equations are satisfied by the function $f_n(z,\theta)$ in the series in (6.223)

$$\begin{aligned} &ae^{z}\frac{\partial^2 f_n}{\partial\theta^2} - \frac{\partial f_n}{\partial z} + nae^{z} f_n \\ &= -C[f_{n-1}\frac{\partial f_1}{\partial\theta} + f_{n-2}\frac{\partial f_2}{\partial\theta} + \ldots + f_1\frac{\partial f_{n-1}}{\partial\theta}] . \end{aligned} \tag{6.224}$$

The system of equations (6.224) must give, in the lowest order of δ , a solution identical with (6.220). This will be achieved if we search solutions of (6.224) fulfilling the condition

$$\left(\frac{\partial f}{\partial z}\right)_{z=0} = 0 . \tag{6.225}$$

Putting

$$e^{z} \approx 1 + \delta\zeta , \qquad \frac{\partial f}{\partial z} \approx \delta\zeta\left(\frac{\partial^2 f}{\partial z^2}\right)_{z=0} \tag{6.226}$$

we obtain from (6.224) in the lowest order of δ.

$$\frac{\partial^2 f_n}{\partial\theta^2} + nf_n = -\frac{C}{a}[f_{n-1}\frac{\partial f_1}{\partial\theta} + \ldots + f_1\frac{\partial f_{n-1}}{\partial\theta}] . \tag{6.227}$$

The series solution (6.223) becomes using (6.226):

$$V = \frac{\epsilon}{a} C \mathrm{e}^{-a}[\mathrm{e}^{-a\delta\zeta} \sin\theta + \mathrm{e}^{-a}\mathrm{e}^{-2a\delta\zeta} f_2(\zeta = 0, \theta) + \mathrm{e}^{-2a}\mathrm{e}^{-3a\delta\zeta} f_3(\zeta = 0, \theta) + \ldots](1 + O(\delta)) \,. \quad (6.228)$$

Determination of the amplitude of the asymptotic solution

However, the appropriate solution of (6.227), giving the coefficients in the series (6.228), is already known from our knowledge of Burgers' equation in chapter 5:

$$\frac{\partial U}{\partial x} - U\frac{\partial U}{\partial \vartheta} = \frac{\partial^2 U}{\partial \vartheta^2} \,. \quad (6.229)$$

In order to realize this we first state that the series

$$U(x, \vartheta) = K\mathrm{e}^{-x}[g_1(\vartheta) + \mathrm{e}^{-x} g_2(\vartheta) + \mathrm{e}^{-2x} g_3(\vartheta) + \ldots] \quad (6.230)$$

satisfies (6.229) if $g_n(\vartheta)$ fulfil the system of equations

$$\frac{d^2 g_n}{d\vartheta^2} + n g_n(\vartheta) = -K[g_{n-1}g_1' + g_{n-2}g_2' + \ldots + g_1 g_{n-1}'] \,. \quad (6.231)$$

For $g_1(\vartheta) = \sin\vartheta$ the first term $K\mathrm{e}^{-x}\sin\vartheta$ of the series in (6.230) is a solution of the linear equation

$$\frac{\partial U}{\partial x} = \frac{\partial^2 U}{\partial \vartheta^2} \,. \quad (6.232)$$

In this case the series in (6.230) must be identical with the Fay solution (cf. (5.85)) with $\epsilon = 1$) of (6.229):

$$U(x, \vartheta) = 2\sum_{n=1}^{\infty} \frac{\sin n\vartheta}{\sinh nx} = 4\sum_{n=1}^{\infty} \sin n\vartheta \sum_{\nu=1}^{\infty} \exp[-(1 + 2\nu)nx]$$
$$= 4\sum_{n=1}^{\infty} \mathrm{e}^{-nx} \sum_{\substack{k = 0, 1, \ldots \\ \frac{n}{2k+1} \text{ integer}}}^{2k+1 \leq n} \sin \frac{n\vartheta}{2k + 1} \,. \quad (6.233)$$

Identification of (6.233) with (6.230) gives all $g_n(\theta)$ and the value of K: $K = 4$. On the other hand, solving the system (6.231) with

$g_1(\vartheta) = \sin\vartheta$ without using its connection with Burgers' equation is cumbersome. However, it is easy to see from (6.233) that the solution of (6.231) with $g_1(\vartheta) = \sin\vartheta$ is

$$g_n(\vartheta) = \left(\frac{K}{4}\right)^{n-1} \sum_{\substack{k=0,1,\dots \\ \frac{n}{2k+1}\ \text{integer}}}^{2k+1\leq n} \sin\frac{n\vartheta}{2k+1} . \tag{6.234}$$

In analogy with (6.234) the solution of (6.227) is

$$f_n(\delta\zeta,\theta) = \left(\frac{C}{4a}\right)^{n-1} \sum_{\substack{k=0,1,\dots \\ \frac{n}{2k+1}\ \text{integer}}}^{2k+1\leq n} \sin\frac{n\vartheta}{2k+1} + O(\delta) . \tag{6.235}$$

With (6.235) inserted into (6.228) we see that the series (6.228), for $\epsilon/\delta \ll 1$, is identical with (6.220) if C fulfils the condition

$$C = 4ae^a . \tag{6.236}$$

Thus the asymptotic solution of (6.214) with (6.215) for $z \to \infty$ is (Scott 1981; Enflo 1996)

$$V = 4\epsilon e^a \exp(-ae^z)\sin\theta . \tag{6.237}$$

In physical terms, the problem solved means finding the asymptotic solution of the spherical Burgers' equation (cf. (2.85))

$$\frac{\partial v}{\partial r} + \frac{v}{r} - \frac{\gamma+1}{2c_0^2} v \frac{\partial v}{\partial \tau} = \frac{b}{2c_0^3\rho_0}\frac{\partial^2 v}{\partial \tau^2} \tag{6.238}$$

with the boundary condition at $r = r_0$:

$$v(r_0,\tau) = v_0 \sin\omega\tau . \tag{6.239}$$

Using (6.11)-(6.13), (6.17), (6.19), (6.212) and (6.213) we obtain for the asymptotic wave v_∞:

$$v_\infty = 8ae^a \frac{c_0^2}{r\omega(\gamma+1)} \exp(-a\frac{r}{r_0}) \sin\omega\tau , \tag{6.240}$$

where (cf. (6.213))

$$a = \frac{b\omega^2 r_0}{2c_0^3\rho_0} . \tag{6.241}$$

Comparison with earlier results

An approximate solution for the asymptotic spherical nonlinear sinusoidal wave is given by Shooter, Muir and Blackstock (1974). They neglect the dissipative term in (6.238) for $r < r_{max}$ and the nonlinear term for $r > r_{max}$. The amplitude of the fundamental harmonic $\sin\omega\tau$ decreases like $r^{-1}(1+R_0\ln(\frac{r}{r_0}))^{-1}$ for $r < r_{max}$ and like $r^{-1}\exp(-a\frac{r}{r_0})$ for $r > r_{max}$. The two undetermined parameters, r_{max} and the asymptotic amplitude of $\sin\omega\tau$, are determined by a continuity condition on the amplitude and its r-derivative at $r = r_{max}$. If the method is applied to a plane wave, in which case the exact asymptotic amplitude is known, the ratio e/2 between the approximate and the exact solution is obtained. For a spherical wave the following asymptotic wave, called v_∞^{SMB} is obtained:

$$v_\infty^{SMB} = 4\Gamma(a)\exp(\Gamma(a))\frac{c_0^2}{r\omega(\gamma+1)}\exp(-a\frac{r}{r_0})\sin\omega\tau \ , \tag{6.242}$$

where $\Gamma(a)$ fulfils the equation

$$\Gamma(a) = \frac{1}{\ln\left(\frac{\Gamma(a)}{a}\right)}. \tag{6.243}$$

It is interesting to compare the expressions (6.242) and (6.240), because, as is seen from (6.213), (6.13) and (6.19), for great a-values the spherical wave approaches a plane wave. Putting

$$\frac{\Gamma(a)}{a} = 1+\alpha \ , \qquad \alpha \ll 1 \tag{6.244}$$

we find from (6.243)

$$a = \frac{1-\frac{\alpha}{2}}{\alpha}(1+O(\alpha)) \tag{6.245}$$

$$\Gamma(a) = \frac{1+\frac{\alpha}{2}}{\alpha}(1+O(\alpha)) \tag{6.246}$$

and thus, as expected (Enflo 1996),

$$\frac{v_\infty^{SMB}}{v_\infty} \xrightarrow[a\to\infty]{} \frac{\mathrm{e}}{2} \ . \tag{6.247}$$

For finite a-values we have

$$\frac{v_\infty^{SMB}}{v_\infty} > \frac{\mathrm{e}}{2}\,, \tag{6.248}$$

so that the Shooter-Muir-Blackstock approximation is better for plane than for spherical waves.

6.4.2 Cylindrical periodic waves

Approximate Fay solution

The cylindrical wave equation (6.15), putting $\epsilon^{(3)} = \epsilon$,

$$\frac{\partial V}{\partial R} - V\frac{\partial V}{\partial \theta} = \epsilon\,(\frac{R}{2} + R_0)\frac{\partial^2 V}{\partial \theta^2}\,, \tag{6.249}$$

with the sinusoidal boundary condition at $r = r_0$, i.e. $R = 0$,

$$V(0, \theta) = \sin\theta \tag{6.250}$$

will now be discussed.

For $R \gg 1$ and $\epsilon \ll 1$, we find from (5.124) an approximate Khokhlov-Soluyan solution, making the change $\epsilon \to \epsilon(\frac{R}{2} + R_0)$:

$$V \approx \frac{1}{R+1}[-\theta + \pi\tanh\frac{\pi\theta}{2\epsilon(\frac{R}{2} + R_0)(R+1)}]\,. \tag{6.251}$$

A new variable is defined through

$$R = 2R_0(y-1) \tag{6.252}$$

$$R_0 = \frac{1}{2}d\epsilon^{-\frac{1}{2}}\,, \tag{6.253}$$

where

$$d^2 = \frac{2b\omega^2 r_0}{c_0^3\rho_0} = O(1) \tag{6.254}$$

$$\epsilon = \frac{bc_0}{2\beta^2 v_0^2 r_0\rho_0}\,. \tag{6.255}$$

With the substitution of (6.252)-(6.255) the cylindrical generalized Burgers' equation (6.249) becomes

$$\frac{\partial V}{\partial y} - d\epsilon^{-\frac{1}{2}} V \frac{\partial V}{\partial \theta} = \frac{1}{2} d^2 y \frac{\partial^2 V}{\partial \theta^2} \ . \tag{6.256}$$

The equation (6.256) with the boundary condition (6.250) has an approximate Khokhlov-Soluyan solution

$$V \approx \frac{1}{1 + d\epsilon^{-\frac{1}{2}}(y-1)} [-\theta + \pi \tanh \frac{\pi\theta}{d\epsilon^{\frac{1}{2}}[1 + d\epsilon^{-\frac{1}{2}}(y-1)]y}] \tag{6.257}$$

and an approximate Fay solution

$$V = d\epsilon^{\frac{1}{2}} y \sum_{n=1}^{\infty} \frac{\sin n\theta}{\sinh\{\frac{n}{2} d\epsilon^{\frac{1}{2}}[1 + d\epsilon^{-\frac{1}{2}}(y-1)]y\}} \ . \tag{6.258}$$

Insertion of (6.257) into (6.249) shows that the requirement for (6.247) to be a good approximation is

$$\frac{1}{2\pi \frac{1}{2} d\epsilon^{-\frac{1}{2}}} \frac{1 + 2d\epsilon^{-\frac{1}{2}}(y-1)}{y} \ll 1 \ . \tag{6.259}$$

The inequality (6.259) gives the y-region, for which the approximate solution (6.257) is useful. Since d is $O(1)$ we choose

$$y - 1 = \delta\eta \ , \quad \frac{\epsilon^{\frac{1}{2}}}{\delta} \ll 1 \ , \quad \delta \ll 1 \ . \tag{6.260}$$

The series (6.258) can be written using (6.260):

$$V = 2d\epsilon^{\frac{1}{2}} \sum_{n=1}^{\infty} \sin n\theta \sum_{\nu=0}^{\infty} \exp[-(\frac{1}{2} + \nu) n d\epsilon^{\frac{1}{2}}(1 + d\epsilon^{-\frac{1}{2}}\delta\eta)](1 + O(\delta)) \ . \tag{6.261}$$

Construction of a series solution from the asymptotic solution

By use of the series (6.261) we will now determine the amplitude constant of solution of the asymptotic equation (cf. (6.256))

$$\frac{\partial V}{\partial y} = \frac{1}{2} d^2 y \frac{\partial^2 V}{\partial \theta^2} \ , \tag{6.262}$$

where y is great enough that the nonlinear term in (6.256) can be neglected. The solution of (6.262) shall develop from (6.257) and has the form

$$V = \epsilon^{\frac{1}{2}} C \exp(-\frac{d^2}{4} y^2) \sin\theta . \qquad (6.263)$$

Obviously (6.263) satisfies (6.262) identically. A series solution of (6.256) with the first term given by (6.263) is

$$V = \epsilon^{\frac{1}{2}} C \exp(-\frac{d^2 y^2}{4})[g_1(\theta) + \exp(-\frac{d^2 y^2}{4}) g_2(y,\theta) + \exp(-2\frac{d^2 y^2}{4}) g_3(y,\theta) + \ldots] . \qquad (6.264)$$

Determination of the amplitude of the asymptotic solution

By inserting (6.264) into (6.256) we find that a solution of (6.256) is obtained if the following differential equations are satisfied by the functions $g_n(y,\theta)$ in (6.264):

$$\frac{d^2 y}{2}[\frac{\partial^2 g_n}{\partial \theta^2} + n g_n(y,\theta)] - \frac{\partial g_n}{\partial y} = -dC(g_{n-1}\frac{\partial g_1}{\partial \theta} + g_{n-2}\frac{\partial g_2}{\partial \theta} + \ldots + g_1 \frac{\partial g_{n-1}}{\partial \theta}). \qquad (6.265)$$

The system of equations (6.265) must, in the lowest order of δ, give a solution of (6.256) identical to (6.261). This will be achieved if we search solutions of (6.265) fulfilling the condition

$$\left(\frac{\partial g_n}{\partial y}\right)_{y=1} = 0 . \qquad (6.266)$$

Using (6.260), (6.266) and the one-term Taylor expansion

$$\left(\frac{\partial g_n}{\partial y}\right) = \delta\eta \left(\frac{\partial^2 g_n}{\partial y^2}\right)_{y=1} \qquad (6.267)$$

in (6.265), we obtain in the lowest order of δ:

$$\frac{\partial^2 g_n}{\partial \theta^2} + n g_n = -\frac{2C}{d}\left[g_{n-1}\frac{\partial g_1}{\partial \theta} + g_{n-2}\frac{\partial g_2}{\partial \theta} + \ldots + g_1 \frac{\partial g_{n-1}}{\partial \theta}\right] . \qquad (6.268)$$

The solution to (6.268) can be constructed in the same way as the solution (6.235) to (6.227) for spherical waves. Then a solution (6.264) is obtained, which can be compared with (6.261) in the lowest order of δ. This comparison gives, in the same way that (6.236) is obtained,

$$C = 2d\exp(\frac{d^2}{4}) \; . \tag{6.269}$$

Thus the asymptotic solution of (6.256) with (6.250) for $y \to \infty$ is according to (6.263) and (6.269) (Scott 1981; Enflo 1996)

$$V = 2\epsilon^{\frac{1}{2}} d\exp(\frac{d^2}{4})\exp(-\frac{d^2}{4}y^2)\sin\theta \; . \tag{6.270}$$

In physical terms, the problem solved means finding the asymptotic solution of the cylindrical Burgers' equation (2.85) with $n = 1$. The boundary condition is formally the same as (6.239). Using (6.252), (6.253), (6.254), (6.255) and (6.11), (6.12) in (6.270) we obtain for the asymptotical cylindrical wave v_∞:

$$v_\infty = \frac{c_0^2}{(\gamma+1)\omega(rr_0)^{\frac{1}{2}}} 8\alpha r_0 \exp(\alpha r_0)\exp(-\alpha r)\sin\omega\tau \; , \tag{6.271}$$

where α is given as

$$\alpha = \frac{b\omega^2}{2c_0^3\rho_0} \; . \tag{6.272}$$

Comparison with earlier results

The asymptotic solution v_∞^{SMB} to the cylindrical equation (2.85) for a sinusoidal boundary condition, obtained by the approximate procedure of Shooter, Muir and Blackstock (1974) is

$$v_\infty^{SMB} = v_0\Big(\frac{r_0}{r}\Big)^{\frac{1}{2}} \frac{2}{1+(\gamma+1)\frac{v_0\omega}{c_0^2} r_0^{\frac{1}{2}}(r_{max}^{\frac{1}{2}} - r_0^{\frac{1}{2}})} \exp[-\alpha(r - r_{max})]\sin\omega\tau \; , \tag{6.273}$$

where r_{max} is given by the expression

$$r_{max} = \frac{1}{2\alpha} + \frac{1}{2}r_0\Big(1 - \frac{1}{R_0}\Big)^2\Big(1 + \sqrt{1 + \frac{2R_0^2}{\alpha r_0(R_0-1)^2}}\,\Big) \; . \tag{6.274}$$

Because $R_0 = O(\epsilon^{-\frac{1}{2}})$ (cf. (6.253)) we can write (6.273):

$$v_\infty^{SMB} =$$
$$\left(\frac{r_0}{r}\right)^{\frac{1}{2}} \frac{2c_0^2 \exp[-\alpha(r - r_{max})]}{c_0^2 + (\gamma+1)v_0\omega r_0} \left(\left(\frac{r_{max}}{r_0}\right)^{\frac{1}{2}} - 1\right)^{-1} \sin\omega\tau(1 + O(\epsilon^{\frac{1}{2}})) \,. \tag{6.275}$$

The cylindrical wave approaches a plane wave for $\alpha r_0 \to \infty$. From (6.274) we find

$$\frac{r_{max}}{r_0} = 1 + \frac{1}{\alpha r_0} + O(\epsilon^{\frac{1}{2}}) + O(\alpha^{-2} r_0^{-2}) \,. \tag{6.276}$$

In the lowest order of $\epsilon^{\frac{1}{2}}$ we find from (6.271) and (6.275) :

$$\frac{v_\infty^{SMB}}{v_\infty} = e^{\alpha r_{max}} \left(\left(\frac{r_{max}}{r_0}\right)^{\frac{1}{2}} - 1\right)^{-1} \frac{1}{4\alpha r_0 e^{\alpha r_0}} \xrightarrow[\alpha r_0 \to \infty]{} \frac{e}{2} \,. \tag{6.277}$$

Thus as in (6.247) the expected result is obtained for a cylindrical wave approaching a plane wave and analogously to (6.248) we have for finite values of αr_0:

$$\frac{v_\infty^{SMB}}{v_\infty} > \frac{e}{2} \,, \tag{6.278}$$

so that the Shooter-Muir-Blackstock (1974) approximation is better for plane than for cylindrical waves.

Related investigations

The case $\epsilon \ll 1$, $\epsilon R_0^2 \ll 1$ has been investigated by a series method by Enflo (1985[a]). In this case the old-age constant, in contrast with (6.269) does not depend on a parameter, but is just a number. The number found by Enflo (1985[a]) is 2 and the number later found by numerical calculations is 1.856 (Hammerton and Crighton 1989) and 1.85 (Sachdev and Nair 1989). The difference is not yet explained.

Chapter 7

Nonlinear bounded sound beams

Bounded sound beams have important applications in underwater acoustics (Novikov, Rudenko and Timoshenko 1987) and in medicine (Averkiou and Cleveland, 1999; Carstensen, Law, McKay and Muir 1980). They are modelled by the KZK equation (Zabolotskaya and Khokhlov 1969; Kuznetsov 1971), derived in section 2.3.3. The KZK equation has been the subject of many studies, a review of which is given by Hamilton (1997). It is clear from this review article that in most of these studies the nonlinear term in the KZK equation is treated as a perturbation (weak nonlinearity). The solving process of the equation then consists of solving a linear equation for each degree of approximation.

For weak nonlinearity the quasilinear theory is applied to the non-dissipative case (KZ equation) by Naze Tjøtta, TenCate and Tjøtta (1991) and by Foda (1996) in search of the evolution of a bifrequency excitation in a rectangular duct of semi-infinite length. Nonlinear interaction of sound beams is studied by the KZK equation, using the quasilinear approximation by Darvennes, Hamilton, Naze Tjøtta and Tjøtta (1991). These studies are extended to include the effects of focusing by Naze Tjøtta, Tjøtta and Vefring (1991).

In the case of strong nonlinearity, some analytical techniques have been used for solving the KZ equation. An asymptotic waveform with shock fronts is given by Makov (1997). A solution in the pre-shock region, based on nonlinear geometrical acoustics and a paraxial approach, which means that the solution is expanded in a series in a transverse variable, is given by Hamilton, Khokhlova and Rudenko (1997).

An analytical technique valid for pulsed sound beams generated by a sound source with a Gaussian dependence of the transverse variable is presented by Frøysa and Coulouvrat (1996). They use a renormalization procedure (Ginsberg 1997) counterbalancing nonuniform terms in a regular perturbation solution. The renormalization technique is employed by Coulouvrat (1991[b]) with weak-shock theory to study Gaussian sound beams.

Numerical techniques have been used extensively for solving the KZ and KZK equation (Bakhvalov, Zhileikin and Zabolotskaya 1987) in the strong nonlinearity case. An early numerical integration of the KZ equation, showing the difference between the nonlinear distortion in a sound beam and the distortion in a plane wave, was made by Bakhvalov, Zhileikin, Zabolotskaya and Khokhlov (1976). Numerical results for the KZK equation, based on a Fourier series expansion of the solution in the nearfield are given by Aanonsen, Barkve, Naze Tjøtta and Tjøtta (1984) and by Hamilton, Naze Tjøtta and Tjøtta (1985). The approach is extended to focused beams by Hart and Hamilton (1988). Numerical results for the KZK equation for a pulse radiated by a circular piston are given by Lee and Hamilton (1995) and by Averkiou and Cleveland (1999). Intersecting beams at small angles are studied numerically for the KZK equation by Cahill and Baker (1999). In order to understand the origin of the harmonics oscillations found they study the quasilinear approximation of the KZ equation by the Fourier decomposition method.

The main solution attempts of the KZK equation (2.84) have thus been direct numerical integration, perturbation teory with nonlinearity considered as a perturbation (for weak nonlinearity) or a Fourier series solution (for strong nonlinearity). The two last approaches result in systems of partial differential equations, which have been truncated and solved numerically.

In accordance with the general idea of this book, in this chapter an analytical technique will be used for studying the nonlinear bounded sound beam by the KZK equation in the strong nonlinearity case. The technique is based on an expansion similar to that of Hamilton, Khokhlova and Rudenko (1997) and generalizes an approach done by Sionoid (1992, 1993).

7.1 The KZK equation

7.1.1 Dimensionless KZK equation

The mathematical description of a nonlinear sound beam, including the dissipative effects, is given by the KZK equation (2.91), rewritten in (7.1),

$$\frac{\partial}{\partial \tau}[\frac{\partial v}{\partial x} - \frac{\beta}{c_0^2} v \frac{\partial v}{\partial \tau} - \frac{b}{2\rho_0 c_0^3}\frac{\partial^2 v}{\partial \tau^2}] = \frac{c_0}{2}(\frac{\partial^2 v}{\partial y^2} + \frac{\partial^2 v}{\partial z^2}) \,. \tag{7.1}$$

A mathematical treatment of the equation is facilitated by writing it in terms of dimensionless variables. To that end we introduce a characteristic velocity amplitude of the medium, v_0, a characteristic radius of the beam, a, and a characteristic frequency, ω. The variables v, y, z, and τ are replaced by dimensionless variables by scaling with v_0, a and ω respectively. For the coordinate x along the beam we choose the same scaling as for nonlinear plane lossless waves. This scaling is done in (4.37). The analysis of section 4.3 shows that a shock appears for $\sigma = 1$, which means that the wave has travelled the distance x_{sh} (shock distance, discontinuity length), defined as

$$x_{sh} = \frac{c_0^2}{\beta\omega v_0} \,. \tag{7.2}$$

Thus our new dimensionless variables are:

$$X - X_0 = \frac{\beta\omega v_0}{c_0^2} x \tag{7.3}$$

(The constant X_0 is introduced for practical reasons, as will be seen in section 7.2.)

$$Y = \frac{y}{a} \tag{7.4}$$

$$Z = \frac{z}{a} \tag{7.5}$$

$$\theta = \omega\tau \tag{7.6}$$

$$V = \frac{v}{v_0} \, . \tag{7.7}$$

After insertion of (7.3)-(7.7) into (7.1) we obtain the dimensionless form of the KZK equation suitable for a boundary condition:

$$\frac{\partial}{\partial\theta}\left(\frac{\partial V}{\partial X} - V\frac{\partial V}{\partial\theta} - \epsilon\frac{\partial^2 V}{\partial\theta^2}\right) = \frac{N}{4}\left(\frac{\partial^2 V}{\partial Y^2} + \frac{\partial^2 V}{\partial Z^2}\right) , \tag{7.8}$$

where the dimensionless parameter ϵ is the same as that defined in (2.77) and the new dimensionless parameter N is defined as

$$N = \frac{2c_0^3}{\beta a^2 \omega^2 v_0} \, . \tag{7.9}$$

In the KZK equation (2.97), suitable for initial value problems, instead of (7.3) we make the substitution

$$X = \frac{\omega\xi}{c_0} \tag{7.10}$$

and instead of (7.6) we substitute

$$\theta = \omega t \frac{v_0 \beta}{c_0} \, . \tag{7.11}$$

With the substitutions (7.4), (7.5) and (7.7) we then obtain from (2.97) the alternative form of the KZK equation suitable for an initial condition

$$\frac{\partial}{\partial X}\left(\frac{\partial V}{\partial\theta} + V\frac{\partial V}{\partial X} - \epsilon\frac{\partial^2 V}{\partial X^2}\right) = -\frac{N}{4}\left(\frac{\partial^2 V}{\partial Y^2} + \frac{\partial^2 V}{\partial Z^2}\right) . \tag{7.12}$$

In addition to (7.8) and (7.12) there are two more ways of writing the KZK equation. By changing the signs of V and θ in the KZK equation (7.8) and the signs of V and X in (7.12), we change the sign of the righthand side of these equations. If we give the $\frac{\partial^2 V}{\partial X \partial\theta}$ term a plus sign and the dissipative term a minus sign in (7.8) and (7.12) there are thus four possibilities of writing each of the KZK equations (7.8) and (7.12): the nonlinear term and the righthand side can have any sign.

In addition to the shock distance x_{sh} defined in (7.2) two other distances of physical interest can be defined. One is the characteristic absorption distance

$$x_{abs} = \frac{2c_0^3\rho_0}{b\omega^2} \,. \tag{7.13}$$

From (2.75) and (2.77) we find that

$$\frac{x}{x_{abs}} = \epsilon\sigma \tag{7.14}$$

and from the leading term in Fay's solution (5.85) we then see that the leading term for large x values decays as $\exp(-\frac{x}{x_{abs}})$ because of dissipative absorption.

The third distance of physical interest is the diffraction length x_{dif}, defined as

$$x_{dif} = \frac{\omega a^2}{2c_0} \,. \tag{7.15}$$

The quantity x_{dif} can be obtained from linear theory. Consider a circular transducer of radius a, whose vibrations emits a sound beam. Having travelled the distance $\sim x_{dif}$, the beam starts to spread spherically, and the wave propagates in a cone with a characteristic aperture angle $\phi \sim a/x_{dif}$ (Novikov, Rudenko and Timoshenko 1987, chapter 4.1).

A new aspect of the physical meaning of the numbers ϵ and N is now shown by writing them as

$$\epsilon = \frac{x_{sh}}{x_{abs}} \tag{7.16}$$

$$N = \frac{x_{sh}}{x_{dif}} \,. \tag{7.17}$$

In the analysis leading to the dimensionless KZK equations (7.8) and (7.12) the assumption

$$N = O(1) \tag{7.18}$$

is made. The coordinate along the beam has to be scaled with the smallest of the characteristic distances x_{sh}, x_{dif} and x_{abs}. Now x_{sh} and x_{dif}, because of (7.17) and 7.18), are of the same order of magnitude, so that the scaling (7.3) is appropriate. The equations (7.8) and (7.12) are valid for strong nonlinearity, manifested by the fact that all terms in (7.8), except possibly the ϵ term, are of the same order of magnitude.

On the other hand, in the case of weak nonlinearity the diffraction length x_{dif} is much smaller than the shock distance x_{sh}, so that we have

$$N \gg 1 . \tag{7.19}$$

Making the scaling (cf. (7.3))

$$X - X_0 = \frac{x}{x_{dif}} = \frac{2c_0 x}{\omega a^2} \tag{7.20}$$

we obtain the dimensionless equation (cf. (7.8))

$$\frac{\partial}{\partial \theta}(N\frac{\partial V}{\partial X} - V\frac{\partial V}{\partial \theta} - \epsilon\frac{\partial^2 V}{\partial \theta^2}) = N(\frac{\partial^2 V}{\partial Y^2} + \frac{\partial^2 V}{\partial Z^2}) . \tag{7.21}$$

With the condition (7.19) the first approximation of the solution of (7.21) is the solution of the equation

$$\frac{\partial^2 V}{\partial \theta \partial X} = \frac{\partial^2 V}{\partial Y^2} + \frac{\partial^2 V}{\partial Z^2} . \tag{7.22}$$

Inserting a solution of the form

$$V(X,Y,Z,\theta) = A(X,Y,Z)e^{-i\Omega\theta} \tag{7.23}$$

into (7.22) we obtain for $A(X,Y,Z)$ the "parabolic equation"

$$-i\Omega\frac{\partial A}{\partial X} = \frac{\partial^2 A}{\partial Y^2} + \frac{\partial^2 A}{\partial Z^2} . \tag{7.24}$$

Using the solution of (7.24) with (7.23) as the first approximation, the solution of (7.21) can be found by successive approximations (Novikov, Rudenko and Timoshenko 1987).

7.1.2 Transformation of the KZK equation to a generalized Burgers' equation

The KZK equation in the form (7.8) or (7.12) has been numerically solved for different boundary or initial value conditions by many authors (Novikov, Rudenko and Timoshenko 1987; Bakhvalov, Zhileikin and Zabolotskaya 1987). Until now there has not been much analytical work on the equation. However, one analytical approach is done by P. Sionoid (1993), who studies the initial value problem for a beam bounded only in the Y direction. The equation of this problem is (7.12) without the Z-derivative. After the substitutions

$$Y = \sqrt{2}y^*N \tag{7.25}$$

we obtain

$$\frac{\partial}{\partial X}(\frac{\partial V}{\partial \theta} + V\frac{\partial V}{\partial X} - \epsilon\frac{\partial^2 V}{\partial X^2}) = -\frac{1}{2}\frac{\partial^2 V}{\partial y^{*2}} \ . \tag{7.26}$$

Substituting

$$\xi = X + \frac{y^{*2}}{2\theta} \tag{7.27}$$

and

$$V(\theta, X, y^*) \equiv G(\theta, \xi(\theta, X, y^*)) \tag{7.28}$$

P. Sionoid (1993) obtains the generalized Burgers' equation for cylindrical waves (cf. (2.87))

$$G_\theta + \frac{1}{2\theta}G + G\,G_\xi - \epsilon\,G_{\xi\xi} = 0 \ . \tag{7.29}$$

This treatment is suitable if the wave function V has a prescribed value for $\theta = 1$ at every parabolic surface $\xi = constant$. In this case the solution of (7.29) gives the evolution of the beam according to (7.26).

7.1.3 Expansion of the solution around the center of the beam

The Sionoid approach to the KZK equation will now be generalized (Enflo 2000). The change of variables (7.27) and (7.28) indicates the possibility of making the Taylor expansion

$$V(\theta, X, y^*) = G(\theta, \xi = X + \frac{y^{*2}}{2\theta})$$
$$= G(\theta, X) + \frac{y^{*2}}{2\theta} G_\xi(\theta, X) + \frac{1}{2}(\frac{y^{*2}}{2\theta})^2 G_{\xi\xi}(\theta, X) + \dots \; . \quad (7.30)$$

The fact that the coefficients of the Taylor expansion (7.30) are found by solving the generalized Burgers' equation (7.29) indicates the possibility to find a more general solution of (7.26) as an expansion of $W(\theta, X, y)$ about the center $y^* = 0$ of the beam:

$$V(\theta, X, y^*) = V_0(\theta, X) + \frac{y^{*2}}{2\theta} V_1(\theta, X)$$
$$+ \frac{1}{2!}(\frac{y^{*2}}{2\theta})^2 V_2(\theta, X) + \frac{1}{3!}(\frac{y^{*2}}{2\theta})^3 V_3(\theta, X) + \dots \; , \quad (7.31)$$

Inserting (7.31) into (7.26) and putting the coefficients of each power of y^2 equal on both sides we find for the two first equations

$$\frac{\partial}{\partial X}(V_{0\theta} + V_0 V_{0X} - \epsilon V_{0XX}) = -\frac{1}{2\theta} V_1 \quad (7.32)$$
$$\frac{\partial}{\partial X}(V_{1\theta} - \frac{1}{\theta} V_1 + V_0 V_{1X} + V_1 V_{0X} - \epsilon V_{1XX}) = -\frac{3}{2\theta} V_2 \; . \quad (7.33)$$

The continuation of the system of equations (7.32), (7.33) leads to an infinite system of differential equations with an infinite number of unknowns V_0, V_1, $V_2 \dots$. However, one solution of this system of equations is known, namely

$$V_1 = V_{0X} \; , \quad V_2 = V_{0XX} \; , \quad V_3 = V_{0XXX} \cdots \; , \quad (7.34)$$

where V_0 satisfies the equation (7.29) with G changed into V_0 and ξ changed into X.

The solution (7.34), generalized by putting

$$V_1 = \kappa_1 V_{0X} \; , \quad V_2 = \kappa_2 V_{0XX} \; , \quad (7.35)$$

is inserted into the two equations (7.32) and (7.33):

$$\frac{\partial}{\partial X}(V_{0\theta} + \frac{1}{2}\frac{\kappa_1}{\theta} V_0 + V_0 V_{0X} - \epsilon V_{0XX}) = 0 \quad (7.36)$$
$$\frac{\partial}{\partial X}[V_{0X\theta} + \frac{1}{\theta}(\frac{3}{2}\frac{\kappa_2}{\kappa_1} - 1)V_{0X} + V_0 V_{0XX} + V_{0X}^2 - \epsilon V_{0XXX}] = 0 \; . \quad (7.37)$$

The equation (7.37) follows from (7.36) by derivation not only for the choice (7.34) of $\kappa_2 = \kappa_1 = 1$ but in general for

$$\kappa_2 = \frac{2}{3}\kappa_1(1 + \frac{\kappa_1}{2}) \,. \tag{7.38}$$

Thus it is possible to find a class of solutions of the KZK equation (7.26) represented as power series expansions in $(y^{*2}/2\theta)$ with the coefficients known from solutions of the generalized Burgers' equation (7.36) with κ_1 arbitrary. From (2.86) we know that $\kappa_1 = 1$ corresponds to a generalized Burgers' equation for cylindrical waves and that $\kappa_1 = 2$ corresponds to a generalized Burgers' equation for spherical waves.

If we want to find more terms in the expansion (7.31) than those given by (7.35)-(7.37), we find that the coefficient V_3 in the expansion (7.31) must contain not only V_{0XXX} but also a new contribution, giving a nonlinear term on the righthand side of (7.26). Because of this new contribution the equation for V_3 is in agreement with (7.36) and (7.37).

We write the expansion (7.31), using (7.35) and (7.38):

$$\begin{aligned} V(\theta, X, y) = {} & V_0(\theta, X) + \kappa_1 \frac{y^{*2}}{2\theta} V_{0X}(\theta, X) + \kappa_2 \frac{1}{2}\frac{y^{*4}}{4\theta^2} V_{0XX}(\theta, X) \\ & + \kappa_3 \frac{1}{3!}\frac{y^{*6}}{8\theta^3} V_{0XXX}(\theta, X) + \lambda_3 \frac{y^{*6}}{8\theta^2}(V_{0X}V_{0XX})_X \ldots \,. \end{aligned} \tag{7.39}$$

Inserting (7.39) into (7.26) and putting the coefficients of each power of y^{*2} equal on both sides we find, in addition to (7.36) and (7.37) with (7.38), the equation

$$\begin{aligned} \frac{\partial}{\partial X}\Big(& V_{0XX\theta} - \frac{2}{\theta} V_{0XX} + V_0 V_{0XXX} + 2\frac{\kappa_1^2}{\kappa_2} V_{0X} V_{0XX} + V_{0X} V_{0XX} \\ & - \epsilon V_{0XXX} + \frac{\kappa_3}{\kappa_2}\frac{6 \cdot 5}{2 \cdot 3!} V_{0XX} + \frac{\lambda_3}{\kappa_2}\frac{6 \cdot 5}{2} V_{0X} V_{0XX}\Big) = 0 \,. \end{aligned} \tag{7.40}$$

Requiring that (7.40) is the X-derivative of (7.37) we find, using (7.38)

$$\kappa_3 = \frac{2}{3} \cdot \frac{2}{5}\kappa_1(1 + \frac{\kappa_1}{2})(2 + \frac{\kappa_1}{2}) \tag{7.41}$$

$$\lambda_3 = -\frac{4}{3 \cdot 15}\kappa_1(\kappa_1 - 1) \,. \tag{7.42}$$

In a similar way the coefficient $V_4(\theta, X)$ can be calculated. We summarize:
The nonlinear beam equation

$$\frac{\partial}{\partial X}(V_\theta + VV_X - \epsilon V_{XX}) = -\frac{1}{2}\frac{\partial^2}{\partial y^{*2}}V \tag{7.43}$$

has the power series solution

$$\begin{aligned}
&V(\theta, X, y)\\
&= V_0(\theta, X) + \kappa\frac{y^{*2}}{2\theta}V_{0X}(\theta, X) + \frac{1}{2!}\frac{2}{3}\kappa(1+\frac{\kappa}{2})\frac{y^{*4}}{4\theta^2}V_{0XX}(\theta, X)\\
&+ \frac{1}{3!}\kappa y^{*6}[\frac{2^2}{5!!}(1+\frac{\kappa}{2})(2+\frac{\kappa}{2})\frac{1}{8\theta^3}V_{0XXX}(\theta, X)\\
&\quad - \frac{2^3}{5!!}(\kappa-1)\frac{1}{8\theta^2}(V_{0X}V_{0XX})_X]\\
&+ \frac{1}{4!}\kappa y^{*8}\Big(\frac{2^3}{7!!}(1+\frac{\kappa}{2})(2+\frac{\kappa}{2})(3+\frac{\kappa}{2})\frac{1}{16\theta^4}V_{0XXXX}(\theta, X)\\
&\quad + \frac{2^4}{7!!}(\kappa-1)\frac{1}{16\theta^2}\{(V_{0X}V_{0XX})_{XX\theta}\\
&\quad - (3\kappa+2)\frac{1}{2\theta}(V_{0X}V_{0XX})_{XX} + [V_0(V_{0X}V_{0XX})_X]_{XX}\\
&\quad - \epsilon(V_{0X}V_{0XX})_{XXXX}\}\Big) + \dots ,
\end{aligned} \tag{7.44}$$

where $V_0(\theta, X)$ is given by the generalized Burgers' equation

$$V_{0\theta} + \frac{\kappa}{2\theta}V_0 + V_0V_{0X} - \epsilon V_{0XX} = f(\theta) , \tag{7.45}$$

where $f(\theta)$ is arbitrary.

7.1.4 Solution for a circular beam

A solution similar to (7.44) will also be given for the KZK equation describing a circular symmetric beam. In this case we put

$$Y^2 + Z^2 = R^2 = \frac{1}{2}Nr^2 \tag{7.46}$$

and

$$\frac{\partial^2}{\partial Y^2}+\frac{\partial^2}{\partial Z^2}=\frac{\partial^2}{\partial R^2}+\frac{1}{R}\frac{\partial}{\partial R}=\frac{2}{N}\left(\frac{\partial^2}{\partial r^2}+\frac{1}{r}\frac{\partial}{\partial r}\right). \qquad (7.47)$$

Using (7.47) the KZK equation (7.12) becomes

$$\frac{\partial}{\partial X}\left(\frac{\partial V}{\partial \theta}+V\frac{\partial V}{\partial X}-\epsilon\frac{\partial^2 V}{\partial X^2}\right)=-\frac{1}{2}\left(\frac{\partial^2 V}{\partial r^2}+\frac{1}{r}\frac{\partial V}{\partial r}\right). \qquad (7.48)$$

The circular symmetric beam equation (7.48) has the power series solution

$$\begin{aligned}
&V(\theta,X,r)\\
&=V_0(\theta,X)+k\frac{r^2}{2\theta}V_{0X}(\theta,X)+\frac{1}{2!}\frac{k(k+1)}{2}\frac{r^4}{4\theta^2}V_{0XX}(\theta,X)\\
&\quad+\frac{1}{3!}kr^6\Big[\frac{(k+1)(k+2)}{3!}\frac{1}{8\theta^3}V_{0XXX}(\theta,X)\\
&\qquad-\frac{k-1}{3}\frac{1}{8\theta^2}(V_{0X}V_{0XX})_X\Big]\\
&\quad+\frac{1}{4!}kr^8\Big(\frac{(k+1)(k+2)(k+3)}{4!}\frac{1}{16\theta^4}V_{0XX\theta}(\theta,X)\\
&\qquad+\frac{k-1}{12}\frac{1}{16\theta^2}\{(V_{0X}V_{0XX})_{XX\theta}-\frac{3k+5}{\theta}(V_{0X}V_{0XX})_{XX}\\
&\qquad+[V_0(V_{0X}V_{0XX})_X]_{XX}-\epsilon(V_{0X}V_{0XX})_{XXXX}\}\Big)\ldots. \qquad (7.49)
\end{aligned}$$

V_0 is given as a solution of the equation

$$V_{0\theta}+\frac{k}{\theta}V_0+V_0V_{0X}-\epsilon V_{0XX}=f(\theta), \qquad (7.50)$$

where $f(\theta)$ is arbitrary. The equations (7.44) and (7.49) are suitable when initial conditions are given for the wave function V.

The expansion of the solution to the plane beam equation in the boundary condition form, obtained from (7.8),

$$\frac{\partial}{\partial \theta}\left(\frac{\partial V}{\partial X}-V\frac{\partial V}{\partial \theta}-\epsilon\frac{\partial^2 V}{\partial \theta^2}\right)=\frac{1}{2}\frac{\partial^2}{\partial y^2}V, \qquad (7.51)$$

is obtained from (7.44) by the changes

$$\kappa \to -\kappa\,, \quad \frac{\partial}{\partial X} \to -\frac{\partial}{\partial X}\,, \quad \theta \leftrightarrow X\,. \tag{7.52}$$

Analogously, the expansion of the solution to the circular beam equation in the boundary condition form

$$\frac{\partial}{\partial\theta}\Big(\frac{\partial V}{\partial X} - V\frac{\partial V}{\partial\theta} - \epsilon\frac{\partial^2 V}{\partial\theta^2}\Big) = \frac{1}{2}\Big(\frac{\partial^2 V}{\partial r^2} + \frac{1}{r}\frac{\partial V}{\partial r}\Big) \tag{7.53}$$

is obtained from (7.49) by the changes

$$k \to -k\,, \quad \frac{\partial}{\partial X} \to -\frac{\partial}{\partial X}\,, \quad \theta \leftrightarrow X\,. \tag{7.54}$$

The boundary condition form of the KZK equation is useful for studying the propagation of sound beams prepared by prescribing a disturbation with a common phase on a curved surface and a prescribed transverse dependence of the amplitude.

7.2 Propagation of a shock wave in a sound beam

7.2.1 Determination of the boundary condition from the series solution

An example of a boundary value problem of practical interest will be given. Using (7.54) in (7.49) we obtain the series solution of (7.53)

$$\begin{aligned}
&V(X,\theta,r^2)\\
&= V_0(X,\theta) + k\frac{r^2}{2X}V_{0\theta}(X,\theta) + \frac{1}{2!}\frac{k(k-1)}{2}\frac{r^4}{4X^2}V_{0\theta\theta}(X,\theta)\\
&\quad + \frac{1}{3!}kr^6\Big[\frac{(k-1)(k-2)}{3!}\frac{1}{8X^3}V_{0\theta\theta\theta}(X,\theta) - \frac{k+1}{3}\frac{1}{8X^2}(V_{0\theta}V_{0\theta\theta})_\theta\Big]\\
&\quad + \frac{1}{4!}kr^8\Big(\frac{(k-1)(k-2)(k-3)}{4!}\frac{1}{16X^4}V_{0\theta\theta\theta}(X,\theta)\\
&\qquad + \frac{k+1}{12}\frac{1}{16X^2}\{-(V_{0\theta}V_{0\theta\theta})_{\theta\theta X} - (3k-5)\frac{1}{X}(V_{0\theta}V_{0\theta\theta})_{\theta\theta}\\
&\qquad + [V_0(V_{0\theta}V_{0\theta\theta})_\theta]_{\theta\theta} + \epsilon(V_{0\theta}V_{0\theta\theta})_{\theta\theta\theta\theta}\}\Big)\ldots\,,
\end{aligned} \tag{7.55}$$

where $V_0(X, \theta)$ solves the equation

$$V_{0X} - k\frac{V_0}{X} - V_0 V_{0\theta} - \epsilon V_{0\theta\theta} = f(X) , \qquad (7.56)$$

where $f(X)$ is chosen to be zero.

For the special cases $k = -1$, $k = -1/2$ and $k = 0$, the generalized Burgers' equation (7.56) describes spherical, cylindrical and plane waves respectively (cf. (2.85) and (2.76)). In the present case the equation (7.56) has a formal meaning and has nothing to do with physical spherical, cylindrical or plane waves.

We also note that, for the case $k = 1$, the equation (7.55) becomes

$$V(X, \theta, r^2) = V_0(X, \theta) + \frac{r^2}{2X} V_{0\theta}(X, \theta) - \frac{1}{72}\frac{r^6}{X^2}(V_{0\theta} V_{0\theta\theta})_\theta + \ldots \qquad (7.57)$$

so that the first two terms in the series may be a useful approximation of the total solution

Now we will see from which boundary condition a solution of the form (7.55) can develop. To that end we use physical variables and write (7.1) for a circular beam:

$$\frac{\partial}{\partial \tau}\left(\frac{\partial v}{\partial x} - \frac{\gamma+1}{2c_0^2} v \frac{\partial v}{\partial \tau} - \frac{b}{2c_0^3 \rho_0}\frac{\partial^2 v}{\partial \tau^2}\right) = \frac{c_0}{2}\left(\frac{\partial^2 v}{\partial r_\perp^2} + \frac{1}{r_\perp}\frac{\partial v}{\partial r_\perp}\right) , \qquad (7.58)$$

where (cf. (7.4), (7.5) and (7.46))

$$r_\perp^2 = y^2 + z^2 = a^2 \frac{N}{2} r^2 . \qquad (7.59)$$

The beam is started by a time-dependent disturbance with the same phase on a spherical boundary with radius d. The equation of the boundary surface is

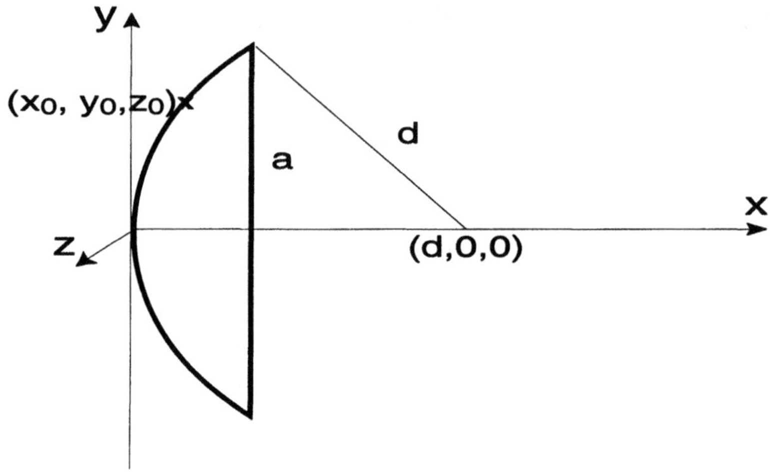

Figure 7.1: The spherical cap on which the beam is started.

$$(x-d)^2 + y^2 + z^2 = d^2 \ . \tag{7.60}$$

or, with $y \ll d$, $z \ll d$:

$$x \approx \frac{y^2+z^2}{2d}. \tag{7.61}$$

Instead of formulating a boundary condition on the curved surface (7.61) we formulate an equivalent boundary condition (Ystad and Berntsen 1995) on the plane $x = 0$. This is easily done by realizing that the phase ωt of the wave at (x_0, y_0, z_0) on the curved surface corresponds to the phase $\omega t + \frac{\omega}{c_0} x_0$ at the plane $x = 0$. Using (7.61) an equivalent boundary condition for v in (7.58) on the plane $x = 0$ can now be written

$$\begin{aligned} v(x=0, r_\perp, t) &= v_0 g(\frac{r_\perp^2}{a^2}) F(\omega t + \frac{1}{2}\frac{\omega r_\perp^2}{c_0 d}) \ , \qquad r_\perp < a \\ v(x=0, r_\perp, t) &= 0 \ , \qquad r_\perp > a \ , \end{aligned} \tag{7.62}$$

where a is the equivalent plane source radius. Using (7.3)-(7.7) and (7.46) in (7.58) we obtain (7.53) with the boundary condition (use (7.59))

$$V(X = X_0, r, \theta) = g(\frac{1}{2} N r^2) F(\theta + \frac{1}{2}\frac{x_{sh}}{d} r^2) \ , \qquad r < \frac{\sqrt{2}}{N}$$

$$V(X = X_0, r, \theta) = 0 \ , \qquad r > \frac{\sqrt{2}}{N} \ . \tag{7.63}$$

With the notation

$$G(r^2) \equiv g(\frac{1}{2}Nr^2) \tag{7.64}$$

we obtain from (7.63):

$$\begin{aligned} V(X = X_0, r, \theta) &= G(r^2)F(\theta + \frac{1}{2}\frac{x_{sh}}{d}r^2) \\ &= [G(0) + r^2G'(0) + \ldots][F(\theta) + \frac{1}{2}\frac{x_{sh}}{d}r^2F'(\theta)\ldots] \\ &= G(0)F(\theta) + r^2[G'(0)F(\theta) + \frac{1}{2}\frac{x_{sh}}{d}G(0)F'(\theta)] \\ &\quad + \frac{r^4}{2}[G''(0)F(\theta) + \frac{x_{sh}}{d}G'(0)F'(\theta) + \frac{x_{sh}^2}{4d^2}G(0)F''(\theta)] \ldots \ , \\ & \qquad r < \frac{\sqrt{2}}{N} \ . \end{aligned} \tag{7.65}$$

We will require that the functions $G(r^2)$ and $F(\theta)$ have such properties that a series solution of the type given in (7.55) can develop from a given time-dependent and r^2-dependent perturbation on the curved surface (7.60). We impose an N-wave boundary condition, which means that for $X = X_0$ the dependence of V_0 on θ is linear. Thus the derivatives $V_{0\theta\theta}(X_0, \theta)$, $V_{0\theta\theta\theta}(X_0, \theta) \ldots$ vanish and the exact boundary condition at $X = X_0$ of the series solution (7.55) is

$$\begin{aligned} V(X = X_0, r, \theta) &= V_0(X_0, \theta) + \frac{kr^2}{2X_0}V_{0\theta}(X_0, \theta) \ , \qquad r < \frac{\sqrt{2}}{N} \\ V(X = X_0, r, \theta) &= 0 \ , \qquad r > \frac{\sqrt{2}}{N} \ . \end{aligned} \tag{7.66}$$

Comparing (7.65) with (7.66) we can now see how an N-wave boundary condition fulfilling (7.66) can be prepared. Requiring $F(\theta)$ to be linear in θ and $G(r^2)$ to be equal to the constant G we find that the terms of order r^4, r^6, ... cancel in (7.65). The boundary condition (7.65) then becomes

$$V(X = X_0, r, \theta) = G\,[F(\theta) + \frac{1}{2}\frac{x_{sh}}{d}r^2F'(\theta)] \ . \tag{7.67}$$

The identification of (7.66) and (7.67) gives

$$V_0(X_0, \theta) = G\,F(\theta) \tag{7.68}$$

$$X_0 = \frac{kd}{x_{sh}} = \frac{k\beta\omega v_0 d}{c_0^2} \; . \tag{7.69}$$

Assuming that the duration of the original N-wave is $2T$, we can now put $\omega = \frac{1}{T}$. This means that the N-wave in the boundary condition is different from zero in the interval $-1 < \theta < 1$.

7.2.2 Solution of generalized Burgers' equation

The equation (7.56) will now be solved. It can be transformed into a generalized Burgers' equation of standard form by changes of variables:

$$W = -\frac{V_0}{X^k} \tag{7.70}$$

$$\xi = \frac{1}{k+1} X^{k+1} \; . \tag{7.71}$$

Changing the variables V_0 and X into W and ξ in (7.56) we obtain a generalized Burgers' equation for W:

$$W_\xi + W W_\theta - \epsilon[(k+1)\xi]^{-k/(k+1)} W_{\theta\theta} = 0 \; . \tag{7.72}$$

For arbitrary $k \neq -1$ the wave equation (7.72) formally describes plane waves propagating in a medium with variable viscosity.

For $k \leq 0$ the solutions of (7.72) give decaying shocks, as we have seen in the old-age N-wave solutions (5.41) (for plane waves) and (6.94) (for cylindrical waves). On the other hand, for $k > 0$ there are solutions of (7.72) giving preserved shocks. As will be seen in this subsection, $k = 1$ is the lower limit for non-increasing shockwidth during propagation. We put $k = 1$ in (7.72) and obtain

$$W_\xi + W W_\theta - (\epsilon/\sqrt{2\xi}) W_{\theta\theta} = 0 \; , \tag{7.73}$$

which equation will be studied with an N-wave boundary condition. This condition, corresponding to (4.28), (4.29), is

$$W(1, \theta) = \theta \; , \qquad |\theta| < 1 \tag{7.74}$$

$$W(1, \theta) = 0 \; , \qquad |\theta| > 1 \tag{7.75}$$

leading to the outer solution, corresponding to (4.26), (4.27):

$$W^{(e)}(\xi,\theta) = \frac{\theta}{\xi} + o(\epsilon^n) , \quad |\theta| < \sqrt{\xi} \tag{7.76}$$

$$W^{(e)}(\xi,\theta) = 0 , \quad |\theta| > \sqrt{\xi} . \tag{7.77}$$

The inner solution of (7.73), found by matching to (7.76), (7.77), is obtained in the same way as (5.14). The inner solution W^* is:

$$W^* = \frac{1}{2\sqrt{\xi}}[1 - \tanh\frac{\theta^* - \sqrt{2}(1-\xi^{\frac{1}{2}})}{2\sqrt{2}}] + \epsilon W_1^* + O(\epsilon^2) , \tag{7.78}$$

where

$$\theta^* = \frac{\theta - \sqrt{\xi}}{\epsilon} . \tag{7.79}$$

The solution (7.76) - (7.78) is valid until the second term in the asymptotic series (7.78) is of the same order of magnitude as the first one. It can be seen from the calculation of the next term in (7.78) by Crighton and Scott (1979) that this happens for $\xi = O(\epsilon^{-2})$. In order to obtain a solution of (7.72), valid for $\xi = O(\epsilon^{-2})$, we make the new scaling

$$\xi_I = \epsilon^2 \xi . \tag{7.80}$$

The new scalings of θ and W follows from (7.72) and (7.80):

$$\theta_I = \epsilon\theta , \tag{7.81}$$

$$W_I = \epsilon^{-1} W . \tag{7.82}$$

The new generalized Burgers' equation is obtained from (7.72) and (7.80)-(7.82):

$$\frac{\partial W_I}{\partial \xi_I} + W_I \frac{\partial W_I}{\partial \theta_I} - \frac{\epsilon^2}{\sqrt{2\xi_I}} \frac{\partial^2 W_I}{\partial \theta_I^2} = 0 . \tag{7.83}$$

The outer solution of (7.83) is identical with (7.76), (7.77). The inner solution of (7.83) is, written in the same variables as (7.78):

$$\epsilon W_I^* = \frac{1}{2\sqrt{\xi}}[1 - \tanh\frac{(\theta - \sqrt{\xi})/\epsilon - \sqrt{2}(1-\epsilon\xi^{\frac{1}{2}})}{2\sqrt{2}}] + O(\epsilon^2) . \tag{7.84}$$

The solution (7.84) shows that for $\sqrt{\xi} = O(\frac{1}{\epsilon})$ we have a shock solution whose discontinuity is $1/\sqrt{\xi}$. The shock center in the solution (7.78) is at

$$\theta = \theta_C \equiv \sqrt{\xi} + \epsilon\sqrt{2}\left(1 - \sqrt{\xi}\right) , \qquad \xi = O(1) \tag{7.85}$$

and in the solution (7.84) at

$$\theta = \sqrt{\xi} + \epsilon\sqrt{2}\left(1 - \epsilon\sqrt{\xi}\right) , \qquad \xi = O(\epsilon^2) . \tag{7.86}$$

Thus in the solution (7.84), because $\xi = O(\epsilon^{-2})$, it is seen, by comparing (7.85) and (7.86), that the shock has not moved farther from the location $\theta = \sqrt{\xi}$ than it has already moved in the solution (7.78). We also see from (7.78) and (7.84) that the coefficient $\epsilon/\sqrt{2\xi}$ of the second derivative in (7.72) has exactly the necessary decreasing behaviour with ξ for giving the nongrowing shock width $2\sqrt{2}\,\epsilon$. The solution (7.84) is no longer valid for $\xi = O(\epsilon^{-4})$ and a new rescaling similar to that made in the equations (7.80)-(7.82) must be made. For every new rescaling the small parameter is squared (ϵ, ϵ^2, ϵ^4, ...) and the shock center does not move farther from the location $\theta = \sqrt{\xi}$.

We thus conclude that the shock is preserved in solutions of (7.72) with $k \geq 1$.

7.2.3 Conditions for shock preservation

It remains to formulate, in physical variables, the boundary condition by which a beam with a preserved shock is prepared. The function $F(\theta)$ introduced in (7.62) is chosen so that its maximum is equal to 1. The boundary $X = X_0$ corresponds to the boundary $\xi = 1$, which means

$$X_0 = (k+1)^{1/(k+1)}, \quad k \geq 1 \tag{7.87}$$

by use of (7.71). Because the maximum value of W is equal to 1, then by use of (7.87) and (7.70) we find that the maximum value of V_0 is equal to $(k+1)^{k/(k+1)}$, and therefore using (7.68) and (7.64):

$$\begin{aligned} G(r) &= (k+1)^{k/(k+1)} , \qquad r^2 < \frac{2}{N} \\ G(r) &= 0 , \qquad r^2 > \frac{2}{N} . \end{aligned} \tag{7.88}$$

With the notation

$$u_0 = (k+1)^{k/(k+1)}\, v_0 \tag{7.89}$$

and the assumption $k \geq 1$ the boundary condition (7.62) of (7.58) for a beam with a preserved N-wave can be formulated, using (7.70), (7.74), (7.75), (7.88) and (7.89):

$$\begin{aligned} v(x=0, r_\perp, t) &= u_0 F(\omega t + \frac{1}{2}\frac{\omega r_\perp^2}{c_0 d}) \ , \qquad && r_\perp < a \\ v(x=0, r_\perp, t) &= 0 \ , && r_\perp > a \ , \end{aligned} \tag{7.90}$$

where

$$\begin{aligned} F(s) &= -s \ , \qquad && |s| < 1 \\ F(s) &= 0 \ , && |s| > 1 \ . \end{aligned} \tag{7.91}$$

The focal length of the source is given by use of (7.69) and (7.87) as

$$d = \frac{(k+1)c_0^2 T}{k\beta u_0} \ , \tag{7.92}$$

where $2T \equiv 2/\omega$ is the duration of the N-wave pulse, u_0 is the maximum velocity in the pulse, c_0 is the sound velocity and β is given by (2.72) (Enflo 2000). It is also assumed that ϵ, given by (7.2), (7.13) and (7.16), is much less than unity. The condition $k \geq 1$ for preservation of the shock gives, inserted into (7.92):

$$d \leq \frac{2c_0^2 T}{\beta u_0} \ . \tag{7.93}$$

In (7.92) is given the maximal possible focal length of the source cap for the shock to be preserved. For $k < 1$ the term corresponding to the second derivative term in (7.72) does not decrease sufficiently fast for the shock width to be constant and the focal length d becomes larger than the maximum value given in (7.92). An example of a non-preserved shock ($k < 1$) is given in Figure 7.2 and an example of a preserved shock ($k > 1$) is given in Figure 7.3.

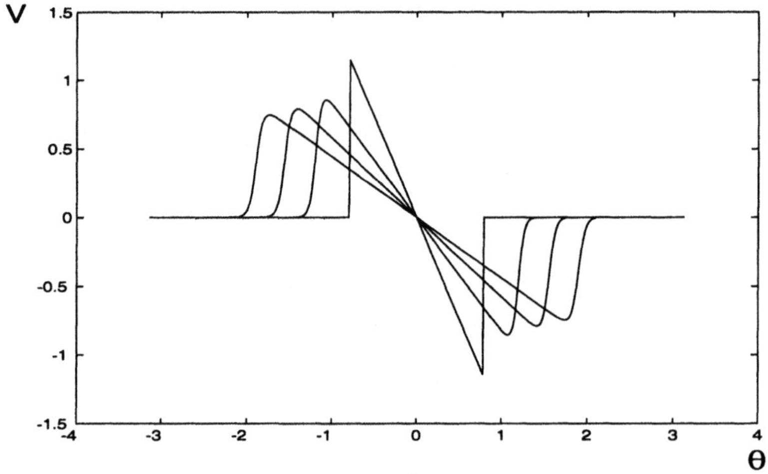

Figure 7.2: From equation (7.56) with $\epsilon = 0.02$, at propagated distances $X - X_0 = 0$, 1, 2 and 3 where $X_0 = 1.3104$ and $k = 0.5 < 1$ i.e. a diverging beam.

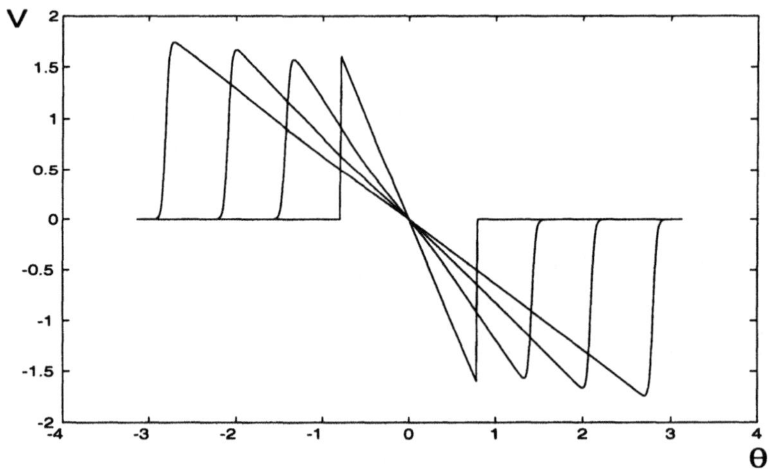

Figure 7.3: From equation (7.56) with $\epsilon = 0.02$, at normalized propagated distances $X - X_0 = 0$, 1, 2 and 3 where $X_0 = 1.4364$ and $k = 1.3 > 1$ i.e. a converging beam.

Chapter 8

Nonlinear standing waves in closed tubes

Nonlinear acoustical theory must be used when the amplitude of the oscillations of the medium are sufficiently great, so that the acoustical Reynolds number, defined in (2.77), is at least of the order of unity. This means that, keeping all other parameters constant, the amplitude of the oscillation can be increased to a level which makes nonlinear acoustic theory necessary to use. In closed tubes this increase of the amplitude can be achieved by choosing the frequency of the oscillation to be a resonance frequency of the tube. Nonlinear oscillations of a gas contained in a tube of finite length have been studied experimentally and theoretically by many authors. For work before 1964 and for gas oscillations in open tubes we refer to the review article by Ilgamov, Zaripov, Galiullin and Repin (1996).

Today finite-amplitude acoustic phenomena in resonant tubes are of special interest for thermoacoustic heat engines (Gaitan and Atchley 1993). The complete problem of nonlinear standing waves in a cavity requires a three-dimensional solution of a nonlinear wave equation with consideration of thermal and viscous losses at the walls of the cavity, modelled by boundary layers. This becomes rather involved and the problem has to be approached by simplifying approximations.

An effective approximate approach for calculating resonant oscillations in closed tubes was presented by Chester (1964). The wall friction is treated by a boundary layer, which is assumed to have a small displacement effect on the main one-dimensional flow. This flow is modelled as a boundary value problem of the fundamental hydrodynamic equations, with one closed and one oscillating end of a tube. Three cases are analyzed; the inviscid case, the case with compressive viscocity and the boundary layer effect.

Chester's work has guided much later experimental and theoretical work. Chester's theoretical results are in good agreement with experimental data for weak shock waves, as shown by Temkin (1968), Cruikshank (1972) and Lawrenson, Lipkins, Lucas, Perkins and Van Doren (1998). Theoretical development of Chester's work is done in the inviscid case by Jimenez (1973) for a variety of boundary conditions from closed to open mouth at the non-oscillating end of the tube, by Keller (1976) for wall friction effects of arbitrary strength, by Van Wijngaarden (1978) for shock solutions with closed and open non-oscillating end, by Ni (1983) for continuous waves and shock waves obtained by successive approximations from the unperturbed gas and by Nyberg (1999) for a bifrequency oscillation at one end of a closed tube. A numerical model for nonlinear standing waves, based on a nonlinear differential equation written in Lagrangian coordinates, is presented by Vanhille and Campos-Pozuelo (2001). The first theoretical work on nonlinear oscillations in tubes starting with Kuznetsov's equation instead of fundamental gas dynamic equations seems to be Gusev (1984), who derives a superposition of counterpropagating simple waves representing standing waves in the inviscid case. Later authors using Kuznetsov's equation for the theoretical modelling of nonlinear oscillations in closed tubes are Coppens and Atchley (1997), who use a perturbation method resulting in successive linear equations, Nyberg (1999), and Ilinskii, Lipkins, Lucas, Van Doren and Zabolotskaya (1998), who consider additional third-order terms in a lossless model, numerically studied.

An early study of finite amplitude standing waves in tubes is made by Coppens and Sanders (1968). They present a perturbation expansion for a second-order one-dimensional nonlinear wave equation. The wall friction is treated by introducing an additional absorption process generating the same energy loss as occurs within the boundary layer. In a later work (Coppens and Sanders 1975) they extend their work to a three-dimensional model for lossy cavities. Different resonator shapes and boundary layer effects are studied by Bednarik and Cervenka (1999).

If the vibration at the end of the tube is generated by finite displacements of its boundary, the boundary condition at the moving end by itself excites nonlinear oscillations, although the medium may be linearly deformed. This problem is studied by Rudenko (1999) and Rudenko and Shanin (2000), who find solutions both for the nonstationary problem and for steady-state oscillations. Rudenko, Hedberg and Enflo (2001) present an exact non-steady-state solution corresponding to a sawtooth-like periodic vibration of the boundary in the case where the medium is nonlinearly deformed.

This chapter presents a completion and enlargement of Chester's (1964) theoretical model for the case when the wall friction is neglected. The starting point is Kuznetsov's (1971) equation. It is shown how Kuznetsov's equation leads to the evolution equation used by Rudenko , Hedberg and Enflo (2001). A new mathematical treatment, in comparison with Chester's (1964), is given for the inviscid solution. For the steady-state solution in a dissipative medium a more elaborate perturbation expansion, in comparison with Chester's, is presented.

8.1 Nonlinear and dissipative effects at non-resonant and resonant driving frequencies

The starting point for the mathematical analysis of nonlinear standing waves is Kuznetsov's equation (2.60), rewritten in (8.1):

$$\frac{\partial^2\Phi}{\partial t^2} - c_0^2\Delta\Phi = \frac{\partial}{\partial t}[(\mathrm{grad}\,\Phi)^2 + \frac{1}{2c_0^2}(\gamma-1)\Big(\frac{\partial\Phi}{\partial t}\Big)^2 + \frac{b}{\rho_0}\Delta\Phi] \,. \quad (8.1)$$

It will be used to describe standing waves in a closed gas-filled tube with a piston oscillating at one end. Let the ends of the tube be at $x = 0$ and $x = L$.

8.1.1 Linear theory of standing waves

For principial reasons we first study this problem within the frame of linear theory. Thus neglecting nonlinear terms in (8.1) and neglecting wall effects we write (8.1) with the velocity $v(x,t)$ (cf. (2.44)) as the dependent variable:

$$\frac{\partial^2 v}{\partial t^2} - c_0^2 \frac{\partial^2 v}{\partial x^2} - \frac{b}{\rho_0} \frac{\partial^3 v}{\partial t \partial x^2} = 0 \ . \tag{8.2}$$

The boundary conditions are assumed to be

$$\begin{aligned} v(0,t) &= 0 \\ v(L,t) &= l\omega \cos \omega t \equiv v_0 \cos \omega t \ . \end{aligned} \tag{8.3}$$

Here l is the maximal displacement of the piston. A solution of the form

$$v(x,t) = U(x) \mathrm{e}^{-i\omega t} \tag{8.4}$$

is sought for. Inserting (8.4) into (8.2) we obtain

$$U'' + \frac{\omega^2}{c_0^2 - i\omega b/\rho_0} U = 0 \ . \tag{8.5}$$

Defining the dissipation parameter δ as

$$\delta \equiv \frac{\omega b}{2\rho_0 c_0^2} \frac{\omega L}{c_0} \ll 1 \tag{8.6}$$

and retaining terms of the order δ^2 we obtain from (8.5)

$$U'' + [\frac{\omega}{c_0}(1 + i\frac{c_0}{\omega L}\delta - \frac{3}{2}\frac{c_0^2}{\omega^2 L^2}\delta^2)]^2 U = 0 \ . \tag{8.7}$$

Solving (8.7) and using the boundary conditions (8.3) we find the following solution, correct to the order δ^2, of (8.2):

$$\begin{aligned} v(x,t) = l\omega / [(1 + \frac{\delta^2}{2}) \sin^2 \frac{\omega' L}{c_0} + \delta^2 \cos^2 \frac{\omega' L}{c_0}] \\ \{[(1 + \frac{\delta^2 x^2}{2L^2}) \sin \frac{\omega' L}{c_0} \sin \frac{\omega' x}{c_0} + \delta^2 \frac{x}{L} \cos \frac{\omega' L}{c_0} \cos \frac{\omega' x}{c_0}] \cos \omega t \\ + \delta[\frac{x}{L} \sin \frac{\omega' L}{c_0} \cos \frac{\omega' x}{c_0} - \cos \frac{\omega' L}{c_0} \sin \frac{\omega' x}{c_0}] \sin \omega t\} \ , \end{aligned} \tag{8.8}$$

where ω' is defined as

$$\omega' \equiv \omega\left(1 - \frac{3}{2}\frac{\delta^2 c_0^2}{\omega^2 L^2}\right) . \tag{8.9}$$

The result (8.8) is valid as long as the denominator in (8.8) is $O(1)$, which is the case when ω' is not in the neighbourhood of $\pi c_0/L$. Before writing (8.2), from which (8.8) follows, we have made the assumption that the nonlinear effects can be neglected in comparison with the dissipative effects. In order to formulate this assumption quantatively we will estimate the nonlinear and dissipative terms on the righthand side of (8.1). We notice that both nonlinear terms are of the same order of magnitude, so we need only discuss one of them.

For the first nonlinear term on the righthand side of (8.1) we find in the present case with one spatial dimension, using (8.8):

$$\begin{aligned}\frac{\partial}{\partial t}(\mathrm{grad}\Phi)^2 &= \frac{\partial}{\partial t}\left(\frac{\partial \Phi}{\partial x}\right)^2 = \frac{\partial}{\partial t}(v^2) \\ &\approx \frac{\partial}{\partial t}(l^2\omega^2 \cos^2 \omega t) = \frac{\partial}{\partial t}\left(l^2\omega^2 \frac{1+\cos 2\omega t}{2}\right) \sim l^2\omega^3 .\end{aligned} \tag{8.10}$$

The dissipative term on the righthand side of (8.1) is estimated, using (8.8)

$$\frac{b}{\rho_0}\frac{\partial}{\partial t}(\Delta\Phi) = \frac{b}{\rho_0}\frac{\partial}{\partial t}\frac{\partial^2 \Phi}{\partial x^2} = \frac{b}{\rho_0}\frac{\partial^2 v}{\partial t \partial x} \approx \frac{b}{\rho_0}\frac{\omega^2}{c_0} v \approx \frac{bl\omega^3}{\rho_0 c_0} . \tag{8.11}$$

The condition that nonlinear effects can be neglected in comparison with dissipative effects far from resonance, i.e. ω' not in the neighbourhood of $\omega c_0/L$, is now found by requiring the quotient η between the righthand sides of (8.10) and (8.11) to be small:

$$\eta \equiv \frac{\rho_0 l c_0}{b} \ll 1 . \tag{8.12}$$

The number at the lefthand side of (8.12) is connected with the acoustical Reynolds number at non-resonance (cf. (2.77)). It is independent of ω and depends only on the parameters of the fluid and the maximal displacement of the piston at the end of the tube.

For ω-values in the neighbourhood of a resonance frequency, given by (cf. (8.8))

$$\omega' = N\frac{\pi c_0}{L} , \quad N \text{ integer} , \tag{8.13}$$

resonance is obtained. For the inviscid case the resonance frequencies are $N\pi c_0/L$. For nonzero dissipation the combination of (8.9) and (8.13) gives a correction to the resonance frequencies, which we call ω_N (δ defined in (8.6) is assumed to be small for the appropriate frequency):

$$\omega_N = N\frac{\pi c_0}{L}(1 + \frac{3}{8}\frac{b^2 N^2 \pi^2}{\rho_0^2 c_0^2 L^2} + O(\delta^4)) . \tag{8.14}$$

8.1.2 Discussion of the small numbers in the problem of nonlinear standing waves

The problem of nonlinear standing waves at resonance frequencies was first treated by Chester (1964). Resonance occurs for those frequencies $\omega \approx \omega_N$ for which $\sin \omega L/c_0$ is $O(\delta)$. (The difference betweeen ω' and ω is $O(\delta^2)$ (cf. (8.9)) and can be neglected here). Our investigation of this problem starts with the statement that the displacements of the fluid elements, according to (8.8), increase with a factor δ^{-1} for $\omega \approx \omega_N$. Then the comparison between nonlinearity and dissipation, which was done in (8.10)-(8.12), may no longer give the result that nonlinearity is negligible in comparison with dissipation. The rest of this chapter is devoted to cases where nonlinearity cannot be neglected.

Before our study of nonlinear standing waves we will discuss the small numbers of the problem and relations among them. We have already defined the number δ, connected with dissipation, in (8.6) and the number η, related to the acoustical Reynolds number at nonresonance, in (8.12). The Mach number $\mathbf{M}$ is defined as the ratio between the maximal fluid velocity $l\omega$ at the piston and the sound velocity in the fluid,

$$\mathbf{M} \equiv \frac{l\omega}{c_0} \ll 1 . \tag{8.15}$$

The ratio between the maximal fluid deplacement at the piston and the pipe length, l/L, can be expressed in δ, η, and $\mathbf{M}$, using (8.6),

(8.12) and (8.15):

$$\frac{l}{L} = \frac{\mathbf{M}^2}{2\delta\eta} \, . \tag{8.16}$$

In the case where the piston frequency is close to a resonance frequency ω_N, the number Δ, defined as a measure of the relative difference between the frequencies is small:

$$\Delta \equiv \pi \frac{\omega - \omega_N}{\omega_1} \ll 1 \, . \tag{8.17}$$

where ω_N is defined in (8.14). For ω in the neighbourhood of a resonance frequency,

$$\omega = \frac{N\pi c_0}{L} + \Delta \frac{c_0}{L} \tag{8.18}$$

we find from (8.6), (8.12) and (8.18),

$$\frac{l}{L} = \frac{2\delta\eta}{N^2\pi^2(1+\frac{\Delta}{N\pi})^2} \tag{8.19}$$

and from (8.15) and (8.18),

$$\mathbf{M} = 2\frac{\delta\eta}{N\pi(1+\frac{\Delta}{N\pi})} \, . \tag{8.20}$$

Now we will compare the importance of nonlinearity and dissipation for ω in the neighbourhood of a resonance frequency in the same manner as was used for a non-resonant ω in (8.10)-(8.12). The difference is that the displacement of the fluid according to (8.8) is enhanced by a factor δ^{-1}, so that[1] instead of (8.12), we obtain for $\omega \approx \omega_N$:

$$\left|\frac{\partial}{\partial t}(\text{grad } \Phi)^2\right| \Big/ \left|\frac{b}{\rho_0}\frac{\partial}{\partial t}(\Delta\Phi)\right| \approx \frac{\rho_0 l c_0}{b\delta} = \frac{\eta}{\delta} \, . \tag{8.21}$$

In order to obtain the result that nonlinear effects are more important than dissipative effects at resonance we assume for the small numbers η and δ:

$$\frac{\eta}{\delta} \gg 1 \, . \tag{8.22}$$

[1]As will be shown later in this chapter, the real enhancement is $(\delta\eta)^{-\frac{1}{2}}$. The linear theory, which is not valid at resonance, gives too high enhancement. The difference does not change the conclusion.

Because of this assumption nonlinear acoustical theory must be used for the description of the standing wave; the solution (8.8) is not valid for $\omega \approx \omega_N$ and must be replaced by a solution containing higher harmonics.

8.2 Equations of nonlinear standing waves

8.2.1 Perturbation solution and boundary conditions of Kuznetsov's equation

How shall nonlinear effects at resonance be taken into account in a consistent way ? The appropriate model is Kuznetsov's equation (8.1) with one space dimension:

$$\frac{\partial^2\Phi}{\partial t^2} - c_0^2\frac{\partial^2\Phi}{\partial x^2} = \frac{\partial}{\partial t}\Big[\Big(\frac{\partial\Phi}{\partial x}\Big)^2 + \frac{1}{2c_0^2}(\gamma-1)\Big(\frac{\partial\Phi}{\partial t}\Big)^2 + \frac{b}{\rho_0}\frac{\partial^2\Phi}{\partial x^2}\Big] , \quad (8.23)$$

where

$$\frac{\partial\Phi}{\partial x} = -v_x \equiv -v \ . \quad (8.24)$$

We will investigate the conditions under which the righthand side of (8.23) can be considered as a perturbation at resonance. Since we have assumed that nonlinearity is more important than dissipation at resonance, it is sufficient to estimate the ratio between one of the nonlinear terms of the righthand side and one of the linear terms on the lefthand side of (8.23). For a term of the latter type we make the estimation, using (8.10) and (8.8):

$$\Big|c_0^2\frac{\partial^2\Phi}{\partial x^2}\Big| = c_0^2\Big|\frac{\partial v}{\partial x}\Big| \sim c_0^2\frac{\omega}{c_0}v_0 = v_0c_0\omega \ . \quad (8.25)$$

Thus using (8.10) we obtain for ω far from ω_N:

$$\Big|\frac{\partial}{\partial t}\Big(\frac{\partial\Phi}{\partial x}\Big)^2\Big| \Big/ \Big|c_0^2\frac{\partial^2\Phi}{\partial x^2}\Big| \sim \frac{l^2\omega^3}{v_0c_0\omega} \sim \frac{l\omega}{c_0} \ . \quad (8.26)$$

For $\omega \approx \omega_N$ the ratio (8.26) has to be multiplied by δ^{-1} and becomes using (8.14) and (8.19):

$$\left|\frac{\partial}{\partial t}\Big(\frac{\partial\Phi}{\partial x}\Big)^2\right| \Big/ \left|c_0^2\frac{\partial^2\Phi}{\partial x^2}\right| \sim \frac{l\omega_N}{c_0\delta} \approx \frac{2}{N\pi}\eta \ll 1 \,. \tag{8.27}$$

(The situation of the approximation is still better, since, as will be shown later, the ratio (8.26) is enhanced by $(\delta\eta)^{-\frac{1}{2}}$ at resonance.) Thus we may attempt to solve (8.23) for the problem of nonlinear standing waves at resonance frequencies in a closed tube by a perturbation procedure starting with a solution of (8.23) without the righthand side. A first approximation, neglecting the righthand side of (8.23), of the solution of (8.23) is

$$\Phi(x,t) = \Phi_1(x,t) \equiv F_1(t - x/c_0) + F_2(t + x/c_0) \,, \tag{8.28}$$

where F_1 and F_2 are arbitrary functions. Inserting the first approximation (8.28) into the righthand side of (8.23), the following inhomogeneous equation is found after some calculation :

$$\begin{aligned}\frac{\partial^2\Phi}{\partial t^2} &- c_0^2\frac{\partial^2\Phi}{\partial x^2} \\ &= \frac{1}{c_0^2}\{(\gamma+1)[f_1(t-x/c_0)f_1'(t-x/c_0) + f_2(t+x/c_0)f_2'(t+x/c_0)] \\ &+ (\gamma-3)[f_1(t-x/c_0)f_2'(t+x/c_0) + f_1'(t-x/c_0)f_2(t+x/c_0)] \\ &+ \frac{b}{\rho_0}[f_1''(t-x/c_0) + f_2''(t+x/c_0)]\} \,,\end{aligned} \tag{8.29}$$

where

$$f_{1,2}(t \mp x/c_0) \equiv F_{1,2}'(t \mp x/c_0) \,. \tag{8.30}$$

In order to integrate the equation (8.29) we make the change of variables

$$\xi = t - x/c_0 \,, \qquad \eta = t + x/c_0 \tag{8.31}$$

and obtain:

$$\begin{aligned}4\frac{\partial^2\Phi}{\partial\eta\partial\xi} &= \frac{1}{c_0^2}\{(\gamma+1)[f_1(\xi)f_1'(\xi) + f_2'(\eta)f_2(\eta)] \\ &+ (\gamma-3)[f_1(\xi)f_2'(\eta) + f_1'(\xi)f_2(\eta)] + \frac{b}{\rho_0}[f_1''(\xi) + f_2''(\eta)]\} \,.\end{aligned} \tag{8.32}$$

The equation (8.32) can now be integrated with respect to η and ξ. With a suitable choice of integration constants we obtain, writing the solution using the original variables:

$$
\begin{aligned}
\Phi(x,t)& \\
&= F_1(t-x/c_0)+F_2(t+x/c_0)\\
&+\frac{1}{4c_0^2}\{(\gamma+1)\frac{x}{c_0}[f_1^2(t-x/c_0)-f_2^2(t+x/c_0)]\\
&+(\gamma-3)[F_1(t-x/c_0)f_2(t+x/c_0)+f_1(t-x/c_0)F_2(t+x/c_0)]\\
&+\frac{2bx}{\rho_0 c_0}[f_1'(t-x/c_0)-f_2'(t+x/c_0)]\}\ ,
\end{aligned}
\tag{8.33}
$$

The boundary conditions are the same as those given in (8.3). The first one of these conditions gives, by use of (8.24):

$$
\frac{\partial \Phi}{\partial x}=0\ , \qquad x=0\ . \tag{8.34}
$$

Using (8.34) in (8.33) we obtain

$$
f_2(t)=f_1(t)\ . \tag{8.35}
$$

Putting

$$
f_1(t)\equiv f(t)=F'(t)\ ,
$$

we obtain the following solution for the fluid velocity:

$$
\begin{aligned}
v(x,t)&\equiv-\frac{\partial\Phi(x,t)}{\partial x}\\
&=\frac{1}{c_0}\{f(t-x/c_0)-f(t+x/c_0)\}\\
&\quad-\frac{\gamma+1}{4c_0^3}\{f^2(t-x/c_0)-f^2(t+x/c_0)\\
&\quad-\frac{2x}{c_0}[f(t-x/c_0)f'(t-x/c_0)+f(t+x/c_0)f'(t+x/c_0)]\}\\
&\quad-\frac{\gamma-3}{4c_0^3}\{F(t-x/c_0)f'(t+x/c_0)-f_1'(t-x/c_0)F(t+x/c_0)\}\\
&\quad-\frac{b}{2\rho_0 c_0^3}\{f'(t-x/c_0)-f'(t+x/c_0)\\
&\quad-\frac{x}{c_0}[f''(t-x/c_0)+f''(t+x/c_0)]\}\ .
\end{aligned}
\tag{8.36}
$$

The unknown function f in (8.36) will be determined by the second boundary condition (8.3). It is clear from (8.36) that the second and the fifth term on the righthand side have the same structure as the first term. Because of that and the fact that the remaining terms three, four and six are second-order terms, term two and five on the righthand side can be omitted in (8.36). The consequent redefinition of the function f will not affect the terms three, four and six. The second boundary condition (8.3) then gives, applied to (8.36):

$$\begin{aligned} & l\omega \sin \omega t \\ & \quad = \frac{1}{c_0}[f(t - L/c_0) - f(t + L/c_0)] \\ & \quad + \frac{\gamma + 1}{2}\frac{L}{c_0^4}[f(t - L/c_0)f'(t - L/c_0) + f(t + L/c_0)f'(t + L/c_0)] \\ & \quad + \frac{bL}{2\rho_0 c_0^4}[f''(t - L/c_0) + f''(t + L/c_0)] \\ & \quad + \frac{\gamma - 3}{4c_0^4}[f'(t - L/c_0)F(t + L/c_0) - f'(t + L/c_0)F(t - L/c_0)] \; . \end{aligned} \tag{8.37}$$

In order to write (8.37) in a dimensionless form we introduce the notations:

$$U(\zeta) = \frac{f(\zeta/\omega)}{c_0^2} \tag{8.38}$$

$$\omega(t - L/c_0) = \zeta_1 \; , \qquad \omega(t + L/c_0) = \zeta_2 \; . \tag{8.39}$$

With (8.39) inserted into (8.37) a dimensionless equation for a standing wave in a dissipative fluid is obtained, using (8.15) and (8.6):

$$\begin{aligned} & \mathbf{M} \sin \omega t \\ & \quad = U(\zeta_1) - U(\zeta_2) + \pi\beta[U(\zeta_1)U'(\zeta_1) \\ & \quad + U(\zeta_2)U'(\zeta_2)] + \delta[U''(\zeta_1) + U''(\zeta_2)] \\ & \quad + \frac{\beta - 2}{2}[U'(\zeta_1)\hat{U}(\zeta_2) - U'(\zeta_2)\hat{U}(\zeta_1)] \; , \end{aligned} \tag{8.40}$$

where $\hat{U}' = U$ and we have used (2.72): $\frac{\gamma+1}{2} = \beta$.

8.2.2 Equations of resonant standing waves

We now assume that the piston vibration frequency ω is in the neighbourhood of the first resonance frequency, which means that N is put equal to 1 in (8.18)

$$\omega = \pi c_0/L + \Delta c_0/L\ , \qquad \Delta \ll 1\ . \tag{8.41}$$

With (8.41) we obtain from (8.39):

$$\zeta_1 = \omega t - \pi - \Delta \tag{8.42}$$

$$\zeta_2 = \omega t + \pi + \Delta \tag{8.43}$$

Because U is small and $U(\zeta_2)$ is close to $U(\zeta_1)$ at resonance, the term with the last parenthesis at the righthand side of (8.40) becomes smaller than the other terms in (8.40) at resonance, i.e. for $\Delta \ll 1$. Far from resonance, on the other hand, this term is of the same order of magnitude as the other terms in (8.40).

The smallness of Δ in (8.41) and (8.43) provides a possibility to replace $U(\zeta_1)$ and $U(\zeta_2)$ at the righthand side of (8.40) by two terms from their series expansion:

$$\begin{aligned} &U(\omega t - \pi - \Delta) - U(\omega t + \pi + \Delta) \\ &\quad \approx U(\omega t - \pi) - U(\omega t + \pi) - \Delta[U(\omega t - \pi) + U(\omega t + \pi)]\ . \end{aligned} \tag{8.44}$$

The energy influx during one period from the vibrating boundary to the resonator must be small in comparison with the accumulated energy. Therefore, and because the lefthand side of (8.40) is a periodic function, the unknown function U must be a quasiperiodic one. The change of U in one period therefore can be approximated:

$$U(\omega t - \pi) - U(\omega t + \pi) \approx -2\pi\mu \frac{\partial U(\omega t + \pi)}{\partial(\mu\omega t)}\ , \tag{8.45}$$

where $\mu \ll 1$ is a small parameter. Using (8.44) and (8.45) the equation (8.40) takes the form (Rudenko, Hedberg and Enflo 2001)

$$\frac{\partial U}{\partial T} + \Delta\frac{\partial U}{\partial \zeta} - \pi\beta U\frac{\partial U}{\partial \zeta} - \delta\frac{\partial^2 U}{\partial \zeta^2} = \frac{\mathbf{M}}{2}\sin\zeta\ , \tag{8.46}$$

where the "slow time"

$$T \equiv \frac{\omega t}{\pi}\,, \tag{8.47}$$

and the "fast time"

$$\zeta = \omega t + \pi\,, \tag{8.48}$$

have been introduced. The equation (8.48) was named "inhomogeneous Burgers equation" (Karabutov and Rudenko 1979). The inhomogeneous Burgers equation has been studied by Rudenko (1974) and by Karabutov, Lapshin and Rudenko (1976). It will be studied for steady-state vibrations in nondissipative and dissipative media in the rest of this chapter.

The establishment of a steady-state field in a resonator is a result of a competition between energy inflow from a vibrating source, and losses caused by nonlinear absorption and linear dissipation. The equilibrium state reached at $T \to \infty$ can be described by the equation, obtained by integration of (8.46) for $\frac{\partial U}{\partial T} = 0$:

$$\delta \frac{dU}{d\zeta} + \pi\beta(U^2 - C^2) - \Delta U = \frac{\mathbf{M}}{2} \cos\zeta\,, \tag{8.49}$$

where C is an arbitrary constant, the physical meaning of which will be discussed later.

8.3 Steady-state resonant vibrations in a non-dissipative medium

In the case of no dissipation we put $\delta = 0$ in (8.49) and obtain the algebraical equation:

$$\frac{\pi\beta}{2}(U^2 - C^2) - \Delta U = \frac{\mathbf{M}}{2} \cos\zeta\,. \tag{8.50}$$

The physical meaning of C^2 is found by integrating (8.49) over one period:

$$\overline{U^2} \equiv \frac{1}{2\pi}\int_{-\pi}^{\pi} U^2\, d\zeta = C^2\,. \tag{8.51}$$

Thus C^2 is the normalized intensity of one of two counterpropagating waves (cf. (8.36)). The mean value of U is assumed to be zero,

$$\overline{U} = 0 \,. \tag{8.52}$$

The solution of the equation (8.50) is

$$U = \frac{\Delta}{\pi\beta} \pm \sqrt{(\frac{\Delta}{\pi\beta})^2 + C^2 + \frac{\mathbf{M}}{\pi\beta}\cos\zeta} \,. \tag{8.53}$$

The sign in (8.53) and the magnitude of C are determined by the condition (8.52) (Chester 1964).

8.3.1 Continuous solution

For small Mach numbers, $\mathbf{M} \ll \Delta^2/\pi\beta$, the solution is given by one of the branches of the solution (8.53), namely by the "-" branch for $\Delta > 0$, and by the "+" branch for $\Delta < 0$:

$$U = -\frac{\mathbf{M}}{2|\Delta|}\mathrm{sgn}\Delta\cos\zeta \,, \qquad C^2 = \overline{U^2} = \frac{\mathbf{M}^2}{8\Delta^2} \ll \frac{\mathbf{M}}{\pi\Delta} \,. \tag{8.54}$$

The inequality in the last member of (8.54) justifies the neglect of C^2 in the derivation from (8.53) of the expression for U in the first equation of (8.54).

For increasing Mach numbers the harmonic sound field in (8.54) is replaced by a deformed field given by (8.53). As long as the condition (8.52) can be satisfied by use of only the negative square root in (8.53), there is no discontinuity in the solution and the shock cannot exist. We will now use (8.52) and (8.53) with the negative root in order to obtain the constant C:

$$\begin{aligned} 2\pi\overline{U} &= \int_{-\pi}^{\pi} U\,d\zeta = 2\frac{\Delta}{\beta} - 2\int_0^{\pi}\sqrt{(\Delta/\pi\beta)^2 + C^2 + (\mathbf{M}/\pi\beta)\cos\zeta}\,d\zeta \\ &= 2\frac{\Delta}{\beta} - 2\int_0^{\pi}\sqrt{(\Delta/\pi\beta)^2 + C^2 + \mathbf{M}/\pi\beta - (\mathbf{M}/\pi\beta)(1-\cos\zeta)}\,d\zeta \\ &= 2\frac{\Delta}{\beta} - 2\sqrt{(\Delta/\pi\beta)^2 + C^2 + \mathbf{M}/\pi\beta} \end{aligned}$$

$$\times \int_0^{\pi} \sqrt{1-(2\mathbf{M})/\{(\pi\beta)[(\Delta/\pi\beta)^2+C^2+\mathbf{M}/\pi\beta]\}\sin^2\frac{\zeta}{2}}\,d\frac{\zeta}{2}\cdot 2$$
$$= 2\frac{\Delta}{\beta} - 4\sqrt{(\Delta/\pi\beta)^2+C^2+\mathbf{M}/\pi\beta}$$
$$\times \int_0^{\pi/2} \sqrt{1-(2\mathbf{M})/\{(\pi\beta)[(\Delta/\pi\beta)^2+C^2+\mathbf{M}/\pi\beta]\}\sin^2\theta}\,d\theta = 0\,. \tag{8.55}$$

The Complete Elliptic Integral of the second kind is defined (Abramowitz and Stegun 1964, p. 590, 17.3.4) as

$$E(m) = \int_0^{\pi/2} \sqrt{1-m\sin^2\theta}\,d\theta\,. \tag{8.56}$$

From (8.55) and (8.56) we find the equation from which C^2 can be found:

$$\frac{\Delta}{2\beta} = \sqrt{(\frac{\Delta}{\pi\beta})^2+C^2+\frac{\mathbf{M}}{\pi\beta}}E(\frac{2\mathbf{M}/(\pi\beta)}{[\Delta/(\pi\beta)]^2+C^2+\mathbf{M}/(\pi\beta)})\,. \tag{8.57}$$

The solution of (8.57) can be found written in a parametric form:

$$\frac{\Delta}{\pi\beta} = \pm\frac{2\sqrt{2}}{\pi}\sqrt{\frac{\mathbf{M}}{\pi\beta}}\frac{E(m)}{\sqrt{m}} \tag{8.58}$$

$$C^2 = \frac{\mathbf{M}}{\pi\beta}[\frac{2}{m}-1-8E^2(m)/(\pi^2 m)]\,. \tag{8.59}$$

From (8.56) follows that the argument m of the elliptic integral E is defined in the region $0 \leq m \leq 1$. The inequality

$$\frac{E(m)}{\sqrt{m}} \geq 1\,, \tag{8.60}$$

with $E(1) = 1$ is also obvious from (8.56). Combining (8.58)-(8.60) we obtain

$$\frac{2\sqrt{2}}{\pi}\sqrt{\frac{\mathbf{M}}{\pi\beta}} \leq \frac{|\Delta|}{\pi\beta} \quad \Rightarrow \quad \mathbf{M} \leq \mathbf{M}_* \equiv \frac{\pi}{8\beta}\Delta^2\,. \tag{8.61}$$

8.3.2 Shock solution

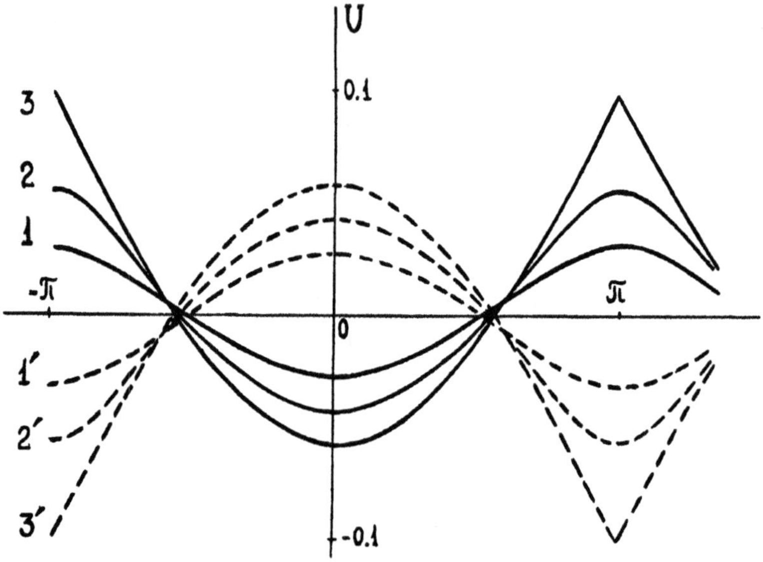

Figure 8.1: Velocity profiles inside the resonator. The solid curves 1, 2 and 3 are constructed for positive frequency discrepancy $\Delta = 0.1\pi\beta$ and the dashed curves for $\Delta = -0.1\pi\beta$. Increase in the curve number corresponds to increase in the amplitude of vibration at the boundary $x = L$: $10^3(\mathbf{M}/\pi\beta) = 5.6, 9.1$ and 12.3.

In Figure 8.1 three U curves are shown for $\Delta > 0$ and $\Delta < 0$ respectively. The curve number 3 corresponds to $\mathbf{M} = \mathbf{M}_*$, at which $\mathbf{M}$-value bifurcations occurs and the steady-state wave form becomes discontinuous. Then a transition from the "+" to the "-" solution and a transition from the "-" to the "+" solution must occur in each period.

However, only a transition which means a compression can be made through a jump, taken from the "-" to the "+" solution. A rarefaction shock, i.e. a jump from the "+" to the "-" solution is prohibited in usual media with quadratic nonlinearity, where the velocity of propagation increases with increasing amplitude of disturbance. In fact, the present situation is analogous to the deformation of a sine wave into a sawtooth wave in a lossless fluid, presented in sections 4.2.3 and 4.2.4. There the compression shock is shown in Figure 4.6. The necessary transition from the "+" to the "-" solution thus must be made without a jump.

For $\mathbf{M} \geq \mathbf{M}_*$ the argument m of the function $E(m)$ has a maximum value of 1. This condition gives directly a condition on C^2, using (8.58) and (8.59):

$$C^2 = \frac{\mathbf{M}}{\pi\beta} - \left(\frac{\Delta}{\pi\beta}\right)^2 . \tag{8.62}$$

Comparing (8.62) with (8.53) we see that the two U curves have a common point for $\zeta = \pm\pi$, in which case the square root in (8.53) equals zero. Thus at $\zeta = \pm\pi$ we have continuous transitions from the "+" to the "-" solution of U, given by (8.53). Inserting (8.62) into (8.53), we obtain the two solutions as

$$U = \frac{\Delta}{\pi\beta} \pm \sqrt{\frac{2\mathbf{M}}{\pi\beta}} \cos\frac{\zeta}{2} . \tag{8.63}$$

The condition (8.52) can now be used to determine the point $\zeta = \zeta_H$, where the jump from the "-" solution to the "+" solution is made:

$$\int_{-\pi}^{\zeta_H} \left(\frac{\Delta}{\pi\beta} - \sqrt{\frac{2\mathbf{M}}{\pi\beta}} \cos\frac{\zeta}{2}\right) d\zeta + \int_{\zeta_H}^{\pi} \left(\frac{\Delta}{\pi\beta} + \sqrt{\frac{2\mathbf{M}}{\pi\beta}} \cos\frac{\zeta}{2}\right) d\zeta = 0 . \tag{8.64}$$

From (8.64) follows

$$\sin\frac{\zeta_H}{2} = \frac{\Delta}{2}\sqrt{\frac{\pi}{2\beta\mathbf{M}}} . \tag{8.65}$$

The result (8.65) is valid for $\mathbf{M} \geq \mathbf{M}_*$ and coincides with $\mathbf{M} = \mathbf{M}_*$ for $\zeta_H = \pm\pi$ for $\Delta > 0$ and $\Delta < 0$ respectively (cf. (8.61)).

The solution (8.63) with a jump from the "-" solution to the "+" solution at $\zeta = \zeta_H$ given by (8.65) is shown in Figure 8.2, which is a continuation of Figure 8.1 for greater Mach numbers.

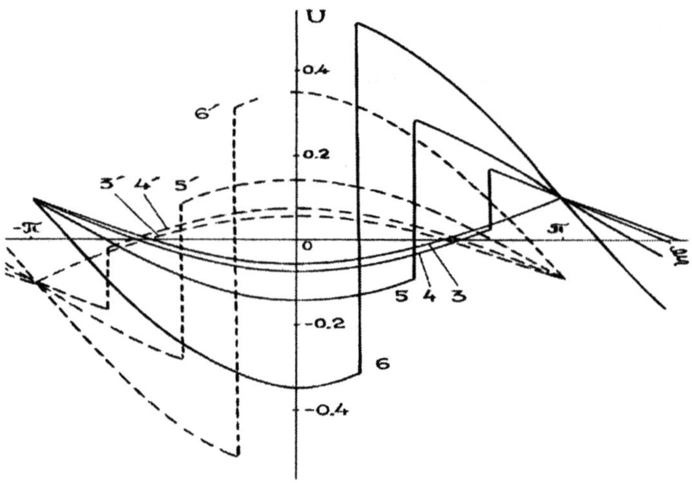

Figure 8.2: Velocity profiles inside the resonator. Curve 3 is the same as curve 3 in Figure 8.1. Curves 4,5 and 6 are for $10^2(\mathbf{M}/\pi\beta) = 1.5$, 3 and 10 where solid ones are for $\Delta = 0.1\pi\beta$ and dashed for $\Delta = -0.1\pi\beta$.

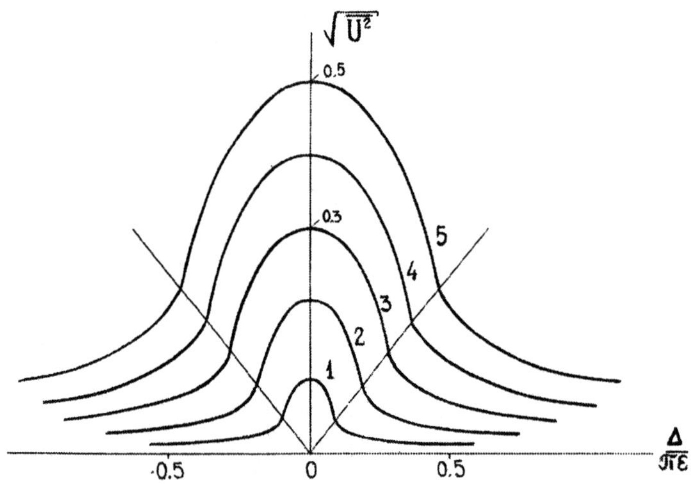

Figure 8.3: Nonlinear frequency response as a function of the discrepancy Δ.

In Figure 8.2 curve 3 corresponds to $\mathbf{M} = \mathbf{M}_*$ and is the same as curve 3 in Figure 8.1. With increasing $\mathbf{M} > \mathbf{M}_*$ the shock appearing initially at $\zeta = \pi$ (for $\Delta > 0$) moves to the position $\zeta = 0$, which can be reached only for $\mathbf{M} \to \infty$. The solid curves 4,5 and 6 are constructed for $\Delta = 0.1\pi\beta$ and $10^2(\mathbf{M}/\pi\beta) = 1.5$, 3 and 10 respectively. The dashed curves demonstrate similar behaviour of the wave profile at $\Delta = -0.1\pi\beta$. In this case, the shock appears for $\mathbf{M} = \mathbf{M}_*$ at $\zeta = -\pi$ and moves to $\zeta = 0$ for $\mathbf{M} \to \infty$.

In Figure 8.3 the nonlinear frequency response in curves 1-5 are constructed at different Mach numbers $10^2(\mathbf{M}/\pi\beta) = 1$, 4, 9, 16 and 25, demonstrating the dependence of $C = \sqrt{\overline{U^2}}$ on the discrepancy (8.17) from the first resonance frequency ω_1. Below the straight lines

$$\sqrt{\overline{U^2}} = \pm\frac{\Delta}{\pi\beta}\sqrt{\pi^2/8 - 1}\,, \tag{8.66}$$

which are obtained by inserting $\mathbf{M} = \mathbf{M}_*$ (cf. (8.61)) into (8.62), the curves are constructed from (8.57), and the wave profile of U has no shocks. Above the straight lines, the equation (8.62) is used for calculation of the frequency response of the discontinuous solution.

8.3.3 The Q-factor

The Q-factor of a resonator can be defined both in the linear and in the nonlinear case. In the linear case, we can use the definition of the Q-factor as the ratio between the amplitudes of the internal and the external (driving) vibrations at resonance ($\Delta = 0$). This ratio is given from formula (8.8) as $(l\omega\delta\sin(\omega x/c_0))/(l\omega\delta^2)$

If we use the square root of the mean value of $\sin^2(\omega x/c_0)$ over the resonator, which is equal to $\sqrt{1/2}$, we obtain for the linear Q-factor Q_L, using (8.6):

$$Q_L = \frac{1}{\delta\sqrt{2}} = \sqrt{2}\frac{\rho_0 c_0^3}{b\omega^2 L} = \frac{\sqrt{2}}{\pi}\frac{\rho_0 c_0^2}{b\omega}\,. \tag{8.67}$$

The same definition of Q as the definition of Q_L above gives for the nonlinear Q-factor Q_{NL}, using (8.62) and (8.15):

$$Q_{NL} = \frac{c_0}{l\omega}(\sqrt{\overline{U^2}})_{\Delta=0} = \frac{c_0}{l\omega}\sqrt{\frac{\mathbf{M}}{\pi\beta}} = 1/\sqrt{\pi\beta\mathbf{M}}\,. \tag{8.68}$$

The Q-factor can also be defined as the ratio between the resonant frequency and the width of the spectral line of the frequency response. This means, using (8.58) with $m = 1$:

$$Q_{NL} = \frac{1}{\Delta} = \frac{\pi}{2\sqrt{2}} \frac{1}{\sqrt{\pi\beta \mathbf{M}}} , \tag{8.69}$$

which differs slightly from (8.68). The nonlinear Q-factor Q_{NL} thus is proportional to $(\beta\mathbf{M})^{-1/2}$ with a coefficient $O(1)$ depending on the definition.

8.4 Steady-state resonant vibrations in a dissipative medium

Now the dissipation number δ is assumed to be nonzero in (8.49), which is written using the new variable $\tau = \zeta/2$:

$$(U - \frac{\Delta}{\pi\beta})^2 + \frac{\delta}{\pi\beta}\frac{dU}{d\tau} = C^2 + \frac{\mathbf{M}}{\pi\beta}\cos 2\tau + \frac{\Delta^2}{\pi^2\beta^2} . \tag{8.70}$$

With other notations this equation is found in Chester's (1964) paper. Following this paper we will present its solution for $\Delta = 0$ in terms of Mathieu functions. For $\Delta \neq 0$ we will present a uniform solution, found by asymptotic matching and not presented in earlier work.

8.4.1 Mathieu equation solution

The equation (8.70) can be linearized, using the transformation

$$U = \frac{\delta}{\pi\beta}\frac{d}{d\tau}\{\ln[we^{\Delta\tau/\delta}]\} . \tag{8.71}$$

Inserting (8.71) into (8.70) we obtain the linear equation

$$\frac{d^2 w}{d\tau^2} - [\frac{\pi^2\beta^2}{\delta^2}(C^2 + \frac{\Delta^2}{\pi^2\beta^2}) + \frac{\mathbf{M}\pi\beta}{\delta^2}\cos 2\tau]w = 0 . \tag{8.72}$$

The equation (8.72) is the Mathieu equation, which we, for future use, write in its canonical form (Abramowitz and Stegun 1964)

$$\frac{d^2w}{d\tau^2} + [a - 2q\cos 2\tau]w = 0 \; . \tag{8.73}$$

The periodic solutions to the equation (8.73) are called Mathieu functions (McLachlan 1947). In order to investigate whether Mathieu functions are useful in the solution of (8.70) we use the condition (8.52), since the mean value of U shall still be zero. From (8.71) we find:

$$\overline{U} \equiv \frac{1}{\pi}\int_{-\pi/2}^{\pi/2} U\, d\tau = \frac{\Delta}{\pi\beta} + \frac{\delta}{\pi\beta}(\ln w(\frac{\pi}{2}) - \ln w(-\frac{\pi}{2})) \; . \tag{8.74}$$

Thus we can use periodic solutions of (8.72) in the case $\Delta = 0$, but not in the case $\Delta \neq 0$.

In the zero discrepancy case, $\Delta = 0$, the Mathieu equation (8.73) will be studied with the parameter values

$$a = -\frac{\pi^2\beta^2}{\delta^2}C^2 \tag{8.75}$$

$$q = \frac{\mathbf{M}\pi\beta}{2\delta^2} \; . \tag{8.76}$$

In order to obtain even periodic solutions of (8.73) the parameter a must have a characteristic value $a_r(q), r = 0, 1, \ldots.$ (Abramowitz and Stegun 1964, chapter 20.1). We are here interested in q values $q \gg 1$. Then the characteristic value $a_0(q)$ has an asymptotic expansion (Abramowitz and Stegun 1964, formula 20.2.30)

$$a_0(q) \sim -2q + 2\sqrt{q} - \frac{1}{4} - \frac{1}{32\sqrt{q}} + \ldots \; . \tag{8.77}$$

The relation (8.51), $\overline{U^2} = C^2$. together with (8.75) and (8.77), now makes an evalution of the Q-factor possible. We obtain, with the definition (8.68):

$$\begin{aligned} Q &= \frac{1}{\mathbf{M}}\sqrt{\overline{U^2}} \\ &= \frac{1}{\mathbf{M}}\frac{\delta}{\pi\beta}\sqrt{-a_0} = \frac{\delta}{\pi\beta\mathbf{M}}\sqrt{2}q^{1/2}[1 - \frac{1}{\sqrt{q}} + \frac{1}{8q} + \ldots]^{1/2} \\ &= \frac{1}{\sqrt{\pi\beta\mathbf{M}}}[1 - \frac{1}{2\sqrt{q}} - \frac{1}{16q} + \ldots] \; , \end{aligned} \tag{8.78}$$

where q is given in (8.76).

The even periodic solution, corresponding to the characteristic value a_0 given by (8.77) is:

$$w(\tau) = ce_0(\tau, q) , \tag{8.79}$$

where $ce_r, r = 0, 1, \ldots$, is a standard notation for even Mathieu functions (Abramowitz and Stegun 1964). With (8.79) and (8.71) the solution to (8.70) for $\Delta = 0$ can be written:

$$U(\tau) = \frac{\delta}{\pi\beta}\frac{d}{d\tau}[\ln(ce_0(\tau, q)] , \quad q = \frac{\mathbf{M}\pi\beta}{2\delta^2} . \tag{8.80}$$

The asymptotic behaviour of $ce_0(\tau, q)$ for $q \gg 1$ is rather complicated; it is found in the book by Abramowitz and Stegun (1964), formula (20.9.11).

8.4.2 Perturbation theory. Matching outer and inner solutions

As mentioned after equation (8.74) we cannot, in the case $\Delta \neq 0$, use the Mathieu functions, i.e. the periodical solutions to the Mathieu equation (8.73), for finding solutions to (8.70). We therefore will attempt at using a perturbation method for solving (8.70) for $\Delta \neq 0$ and the same condition on q as in the preceeding subsection 8.4.1, namely $q \gg 1$.

The equation (8.70) is rewritten using (8.76)

$$(U - \frac{\Delta}{\pi\beta})^2 + \frac{\mathbf{M}^{1/2}}{\sqrt{2\pi\beta q}}\frac{dU}{d\tau} = C^2 + \frac{\mathbf{M}}{\pi\beta}\cos 2\tau + \frac{\Delta^2}{\pi^2\beta^2} . \tag{8.81}$$

Because of (8.51), we have in (8.78) a determination of the constant C^2 in the case $\delta \neq 0$, $\Delta = 0$. On the other hand, in (8.62) the constant C^2 is determined in the case $\delta = 0$, $\Delta \neq 0$. Combining (8.62) and (8.78) we write the two first terms of a series expansion of C^2 in the small parameter $q^{-1/2}$:

$$C^2 = \frac{\mathbf{M}}{\pi\beta} - \frac{\Delta^2}{\pi^2\beta^2} - k\frac{\mathbf{M}}{\pi\beta}q^{-1/2} , \tag{8.82}$$

where k is a constant to be determined. Inserting (8.82) into (8.81), making the substitution

$$U - \frac{\Delta}{\pi\beta} = \sqrt{\frac{2\mathbf{M}}{\pi\beta}}\,V \tag{8.83}$$

and introducing the small parameter ν as

$$\nu = (2q^{\frac{1}{2}})^{-1} \tag{8.84}$$

we obtain

$$V^2 + \nu\frac{dV}{d\tau} = \cos^2\tau - \nu k\ . \tag{8.85}$$

Like the discontinuity at $\zeta = \zeta_H$ (cf. (8.65)) in the inviscid case, the solution of (8.85) has a fast change at $\tau = \tau_0$, where τ_0 will later be determined by the condition $\overline{U} = 0$ (cf. (8.52)). Outside a narrow region around $\tau = \tau_0$, $-\pi/2 < \tau_0 < \pi/2$, the equation (8.85) has an "outer" solution V_{out} expanded in the small parameter ν

$$V_{out} = V_0 + \nu V_1 + \dots\ . \tag{8.86}$$

Inserting (8.86) into (8.85) and putting equal the coefficients of ν^0 and ν on both sides we obtain

$$V_0^2 = \cos^2\tau \tag{8.87}$$

$$2V_0V_1 + \frac{dV_0}{d\tau} = -k \tag{8.88}$$

with the solutions

$$V_0 = \pm\cos\tau \tag{8.89}$$

$$V_1 = \frac{1}{2}(\tan\tau \mp \frac{k}{\cos\tau})\ . \tag{8.90}$$

Like $\sqrt{2\mathbf{M}/(\pi\beta)}V = U - \Delta/\pi\beta$ in (8.63), our solution V_{out}, given by (8.86), (8.89) and (8.90) shall satisfy the conditions (for $\zeta/2 \equiv \tau = \pm\pi/2$)

$$V(\frac{\pi}{2}) = V(-\frac{\pi}{2}) = 0\ . \tag{8.91}$$

A necessary condition for (8.91) to be satisfied by (8.89), (8.90) is obviously

$$k = 1 \tag{8.92}$$

which means agreement, for $\Delta = 0$, between (8.78) and (8.82).

We also search for an inner solution in the neighbourhood of the τ-value $\tau = \tau_0$ at which the outer solution changes (in low order in ν) from $-\cos\tau_0$ to $\cos\tau_0$. (It is shown in subsection 8.3.2 that the change is from the "-" solution to the "+" solution). The inner variable T is introduced as

$$T = \frac{\tau - \tau_0}{\nu} . \tag{8.93}$$

The inner solution V_{in} is expanded in powers of ν:

$$V_{in} = V_0^* + \nu V_1^* + \ldots . \tag{8.94}$$

Inserting (8.94) and (8.93) into (8.85) and putting the coefficients of ν^0 and ν equal on both sides we obtain, putting $k = 1$,

$$\frac{dV_0^*}{dT} = \cos^2\tau_0 - V_0^{*2} \tag{8.95}$$

$$\frac{dV_1^*}{dT} + 2V_0^* V_1^* = -2T\cos\tau_0 \sin\tau_0 - 1 . \tag{8.96}$$

Integration of (8.95) gives

$$V_0^* = \cos\tau_0 \tanh(T\cos\tau_0) , \tag{8.97}$$

where an integration constant has been determined by the requirement

$$V^*(T) = 0 \qquad \text{for } T = 0 . \tag{8.98}$$

The equation (8.96) is solved by means of the integrating factor

$$\exp(2\int V_0^*\, dT) = \cosh^2(T\cos\tau_0) , \tag{8.99}$$

where (8.97) is used. Combining (8.96) and (8.99) we obtain

$$\frac{d}{dT}[V_1^* \cosh^2(T\cos\tau_0)] = -(2T\cos\tau_0\sin\tau_0 + 1)\cosh^2(T\cos\tau_0) . \tag{8.100}$$

Using the indefinite integrals

$$\int \cosh^2 x \, dx = \frac{1}{4} \sinh 2x + \frac{x}{2} \tag{8.101}$$

$$\int x \cosh^2 x \, dx = \frac{x}{4} \sinh 2x - \frac{1}{8} \cosh 2x + \frac{x^2}{4} \tag{8.102}$$

we find from (8.100) the solution to (8.96):

$$\begin{aligned} V_1^* = & \frac{1}{\cosh^2(T\cos\tau_0)}[-\frac{1}{2}T\sin\tau_0\sinh(2T\cos\tau_0) \\ & + \frac{1}{4}\tan\tau_0\cosh(2T\cos\tau_0) - \frac{1}{2}T^2\cos\tau_0\sin\tau_0 \\ & - \frac{1}{4\cos\tau_0}\sinh(2T\cos\tau_0) - \frac{T}{2} - \frac{1}{4}\tan\tau_0] \, , \end{aligned} \tag{8.103}$$

where an integration constant has been determined by the requirement (8.98). The outer and inner solutions are summarized from (8.86), (8.89), (8.90), (8.94), (8.97) and (8.103):

$$V_{out} = \begin{cases} -\cos\tau + \frac{\nu}{2}(\tan\tau + \frac{1}{\cos\tau}) \, , & \tau < \tau_0 \\ \cos\tau + \frac{\nu}{2}(\tan\tau - \frac{1}{\cos\tau}) \, , & \tau > \tau_0 \end{cases} \tag{8.104}$$

$$\begin{aligned} V_{in} = & \cos\tau_0\tanh(T\cos\tau_0) \\ & + \frac{2\nu}{\cosh(2T\cos\tau_0)+1}[-T\sin\tau_0\sinh(T\cos\tau_0)\cosh(T\cos\tau_0) \\ & + \frac{1}{4}\tan\tau_0\cosh(2T\cos\tau_0) - \frac{1}{2}T^2\cos\tau_0\sin\tau_0 \\ & - \frac{1}{4\cos\tau_0}\sinh(2T\cos\tau_0) - \frac{T}{2} - \frac{1}{4}\tan\tau_0] \, . \end{aligned} \tag{8.105}$$

Because of the second term of the righthand side of (8.105), the solutions V_{out} and V_{in} can be matched to each other only if $|\tau_0|$ is not too large. We assume

$$\tau_0 = O(\eta(\nu)) \, , \tag{8.106}$$

where

$$\lim_{\nu\to 0} \eta(\nu) = 0 \tag{8.107}$$

$$\lim_{\nu \to 0} \frac{\eta(\nu)}{\nu} = \infty \tag{8.108}$$

$$\lim_{\nu \to 0} \frac{\eta^2(\nu)}{\nu} = 0 \; . \tag{8.109}$$

Because of (8.106)-(8.109) it is easily seen that, both for $\tau < \tau_0$ and $\tau > \tau_0$, the two expressions for V_{out} and V_{in} in (8.104) and (8.105) respectively attain a common limit for

$$\tau - \tau_0 = O(\eta(\nu)) \; . \tag{8.110}$$

The considerations using (8.106) - (8.110) give an example of the principle of asymptotic matching, previously applied in section 5.1 in this book.

8.4.3 Perturbation theory. Uniform solution

Adding the solutions (8.104) and (8.105) and subtracting the common part, the result is the uniform solution, valid to order ν:

$$\begin{aligned} V = {} & \cos\tau \tanh\left(\frac{\tau - \tau_0}{\nu}\cos\tau_0\right) \\ & + \frac{\nu}{2}\left[\tan\tau \frac{\cosh(2\frac{\tau-\tau_0}{\nu}\cos\tau_0) - 1}{\cosh(2\frac{\tau-\tau_0}{\nu}\cos\tau_0) + 1} - \frac{1}{\cos\tau}\tanh\left(2\frac{\tau - \tau_0}{\nu}\cos\tau_0\right)\right] \\ & - \nu \frac{1}{\cosh(2\frac{\tau-\tau_0}{\nu}\cos\tau_0)}\left[\left(\frac{\tau - \tau_0}{\nu}\right)^2 \cos\tau_0 \sin\tau_0 - \frac{\tau - \tau_0}{\nu}\right] \; . \end{aligned} \tag{8.111}$$

The interval, in which the solution (8.111) is uniform, is somewhat smaller than $(-\frac{\pi}{2}, \frac{\pi}{2})$ because the terms in (8.111) containing $\tan\tau$ and $\cos\tau$ do not compensate each other when τ approaches $\pm\frac{\pi}{2}$ without limit. This reduction of the uniformity interval will be investigated. The value of τ_0 will now be determined using the condition $\overline{U} = 0$ and the integrals

$$\int_{-\pi/2}^{\pi/2} \cos\tau \tanh\left(\frac{\tau - \tau_0}{\nu}\cos\tau_0\right) d\tau = -2\sin\tau_0 + o(\nu) \tag{8.112}$$

$$\int_{-\pi/2}^{\pi/2} \frac{1}{\cos\tau} \tanh\left(\frac{\tau - \tau_0}{\nu}\cos\tau_0\right) d\tau = \ln\frac{1 + \sin\tau_0}{1 - \sin\tau_0} + O(\nu) \; . \tag{8.113}$$

The condition $\overline{U} = 0$ gives, with (8.83) and (8.111)-(8.113)

$$\int_{-\pi/2}^{\pi/2} U \, d\tau$$
$$= \frac{\Delta}{\beta} + \sqrt{\frac{2\mathbf{M}}{\pi\beta}}\left(-2\sin\tau_0 - \frac{\nu}{2}\ln\frac{1+\sin\tau_0}{1-\sin\tau_0}\right) = 0 \ . \quad (8.114)$$

A determination of τ_0 from (8.114) correct to order ν gives (cf. (8.65), where the same determination is made in the case $\nu = 0$)

$$\sin\tau_0 = \frac{\Delta}{2}\sqrt{\frac{\pi}{2\beta\mathbf{M}}} - \frac{\nu}{4}\ln\frac{1+\frac{\Delta}{2}\sqrt{\pi/(2\beta\mathbf{M})}}{1-\frac{\Delta}{2}\sqrt{\pi/(2\beta\mathbf{M})}} \ . \quad (8.115)$$

The mean intensity $\overline{U^2}$ is obtained directly from (8.81) using $\overline{U} = 0$:

$$\overline{U^2} = C^2 - \nu\sqrt{\frac{2\mathbf{M}}{\pi\beta}}\,\overline{\frac{dU}{d\tau}} \ . \quad (8.116)$$

Assuming the conditions (8.91), leading to

$$\overline{\frac{dU}{d\tau}} = 0 \ , \quad (8.117)$$

and using (8.82) with (8.92) we obtain from (8.116):

$$\overline{U^2} = \frac{\mathbf{M}}{\pi\beta} - \frac{\Delta^2}{\pi^2\beta^2} - 2\nu\frac{\mathbf{M}}{\pi\beta} \ . \quad (8.118)$$

It will now be verified that the solution (8.111) gives the result (8.118). Using the integrals

$$\int_{-\pi/2}^{\pi/2} \cos^2\tau \tanh\left(\frac{\tau-\tau_0}{\nu}\cos\tau_0\right) d\tau = \frac{\pi}{2} - 2\nu\cos\tau_0 + O(\nu^2) \quad (8.119)$$

$$\int_{-\pi/2}^{\pi/2} \sin\tau \tanh\left(\frac{\tau-\tau_0}{\nu}\cos\tau_0\right) d\tau = 2\cos\tau_0 + O(\nu) \quad (8.120)$$

$$\int_{-\pi/2}^{\pi/2} \tanh^2\left(\frac{\tau-\tau_0}{\nu}\cos\tau_0\right) d\tau = \pi + O(\nu) \quad (8.121)$$

we obtain directly from (8.111)

$$\overline{V^2} = \frac{1}{\pi}\int_{-\pi/2}^{\pi/2} V^2\,d\tau = \frac{1}{2} - \nu\ , \tag{8.122}$$

which result, together with (8.83), gives (8.118). Thus we have two derivations of (8.118).

The derivations of (8.118) depend on the validity of (8.117) or, expressed in another way, the equality of V in the limits $\tau \to -\pi/2$ and $\tau \to \pi/2$. We will investigate the behaviour of V at these limits and begin with evaluating the limit at $\tau \to \pi/2$ of the critical terms of (8.111). Putting

$$\tau = \frac{\pi}{2} - \mu(\nu)\ , \tag{8.123}$$

where $\mu(\nu)$ has the property

$$\mu(\nu)\underset{\nu\to 0}{\longrightarrow} 0\ , \qquad \mu(\nu) > 0 \tag{8.124}$$

we evaluate the following limit of the coefficient of $\nu/2$ in (8.111):

$$\begin{aligned}
&\lim_{\nu\to 0}\{\tan(\frac{\pi}{2} - \mu(\nu))\frac{\cosh([2(\pi/2 - \tau_0 - \mu(\nu))/\nu]\cos\tau_0) - 1}{\cosh([2(\pi/2 - \tau_0 - \mu(\nu))/\nu]\cos\tau_0) + 1} \\
&\qquad - \frac{\tanh([2(\pi/2 - \tau_0 - \mu(\nu))/\nu]\cos\tau_0)}{\cos(\frac{\pi}{2} - \mu(\nu))}\} \\
&= \lim_{\nu\to 0}\{\frac{1 - \mu^2(\nu)/2}{\mu(\nu)}[1 - 4\exp(-2\frac{\frac{\pi}{2} - \tau_0}{\nu}\cos\tau_0)] \\
&\quad - \frac{1}{\mu(\nu)}[1 - 2\exp(-2\frac{\frac{\pi}{2} - \tau_0}{\nu}\cos\tau_0)]\} \\
&= \lim_{\nu\to 0}\frac{(-2)\exp(-2((\pi/2 - \tau_0)/\nu)\cos\tau_0)}{\mu(\nu)}\ .
\end{aligned} \tag{8.125}$$

The corresponding calculation for

$$\tau = -\frac{\pi}{2} + \lambda(\nu)\ , \tag{8.126}$$

where

$$\lambda(\nu)\underset{\nu\to 0}{\longrightarrow} 0\ , \quad \lambda(\nu) > 0 \tag{8.127}$$

gives

$$\lim_{\nu\to 0}\{\tan(-\frac{\pi}{2}+\lambda(\nu))\frac{\cosh(-2[(\pi/2+\tau_0-\lambda(\nu))/\nu]\cos\tau_0)-1}{\cosh(-2[(\pi/2+\tau_0-\lambda(\nu))/\nu]\cos\tau_0)+1}$$
$$-\frac{\tanh(-2[(\pi/2+\tau_0-\lambda(\nu))/\nu]\cos\tau_0)}{\cos(-\frac{\pi}{2}+\lambda(\nu))}\}$$
$$=\lim_{\nu\to 0}\frac{(-2)\exp(-2[(\pi/2+\tau_0)/\nu]\cos\tau_0)}{\lambda(\nu)}\ . \tag{8.128}$$

We make the following choices for the values of $\lambda(\nu)$ and $\mu(\nu)$:

$$\lambda(\nu)=\mu(\nu)=2\exp(-\frac{c}{\nu})\ ,\quad 0<c<\mathrm{Min}\{\frac{\pi}{2}-\tau_0,\frac{\pi}{2}+\tau_0\}\ . \tag{8.129}$$

From (8.125), (8.128) and (8.129) it follows that $V(\tau)$, given in (8.111), is uniform and finite in the interval

$$-\frac{\pi}{2}+\mu(\nu)<\tau<\frac{\pi}{2}-\mu(\nu)\ . \tag{8.130}$$

Because of (8.125) and (8.128) the value of $V(\tau)$ is $O(\exp(-k/\nu))$, $k>0$, at the ends of the interval (8.130). Thus the assumption (8.117) is proved to be valid. The solution (8.83), (8.111) to (8.81) can be used for all practical purposes, since the impossibility to extend it to exactly $\tau=\pm\pi/2$ has no importance.

8.5 An example of velocity field in a resonator

The velocity field in the resonator is given by (8.36) and is thus found knowing the function $f(\theta)$. Starting with (8.111) we obtain $U(\tau)$ from (8.83). Putting $\tau=\zeta/2$ (cf. (8.70)) and $f(\theta=\zeta/\omega)=c_0^2U(\zeta)$ according to (8.38) we write $f(\theta)$:

$$f(\theta)=c_0^2\sqrt{\frac{2\mathrm{M}}{\pi\beta}}\{\cos\frac{\omega\theta}{2}\tanh(\frac{\omega\theta/2-\tau_0}{\nu}\cos\tau_0)$$
$$+\frac{\nu}{2}[\tan\frac{\omega\theta}{2}\frac{\cosh(2[(\omega\theta/2-\tau_0)/\nu]\cos\tau_0)-1}{\cosh(2[(\omega\theta/2-\tau_0)/\nu]\cos\tau_0)+1}$$

$$
\begin{aligned}
&- \frac{1}{\cos\frac{\omega\theta}{2}} \tanh(2\frac{\omega\theta/2 - \tau_0}{\nu}\cos\tau_0)] \\
&- \nu \frac{1}{\cosh(2[(\omega\theta/2 - \tau_0)/\nu]\cos\tau_0)}[(\frac{\omega\theta/2 - \tau_0}{\nu})^2 \cos\tau_0 \sin\tau_0 \\
&- \frac{\omega\theta/2 - \tau_0}{\nu}]\} + c_0^2\frac{\Delta}{\pi\beta} \,. \qquad (8.131)
\end{aligned}
$$

Putting $\omega = c_0/L(\pi + \Delta)$ in (8.131) according to (8.41) we find, for $\theta = t + x/c_0$ and $\theta = t - x/c_0$, from (8.131) the expressions for $f(\theta)$ which have to be put into (8.36) in order to give the velocity field $v(x,t)$.

Because V is $O(1)$ it follows from (8.83), (8.39) and (8.36) that the ratio $v(x,t)/c_0$ is of the order $\sqrt{\mathbf{M}}$, which according to (8.20), means $O((\delta\eta)^{1/2})$ for $N = 1$. The solution (8.9) of the linear wave equation (8.3), on the other hand, make the ratio $v(x,t)/c_0$ of the order η at resonance frequencies and of the order $\eta\delta$ at non-resonant frequencies. Thus the enhancement factor of the amplitude of the fluid displacement at resonance is of the order $\eta/(\eta\delta) = \delta^{-1}$ according to the linear theory and of the order $(\delta\eta)^{1/2}/(\delta\eta) = (\delta\eta)^{-1/2}$ with due consideration of nonlinearity. Because of (8.22) this means that the enhancement of the fluid displacement amplitude at resonance is lowered by nonlinearity.

An example of an experiment described by the theory accounted for above will be given. The kinematical viscocity b/ρ_0 increases with decreasing air density. We choose, instead of $b/\rho_0 \sim 10^{-4} - 10^{-5}$ m^2/s at atmospheric pressure,

$$
\frac{b}{\rho_0} \approx 10^{-2}\ \mathrm{m^2/s}\ , \qquad (8.132)
$$

and study the lowest resonance frequency ($N = 1$). For the other parameters of the experiment we assume:

$$
\begin{aligned}
L &= 0.5\ \mathrm{m} \\
l &= 10^{-6}\ \mathrm{m} \\
c_0 &= 330\ \mathrm{m/s} \\
\omega_1 &= \frac{\pi c_0}{L} \approx 2000\ \mathrm{s^{-1}}
\end{aligned}
$$

$$\gamma = 1.4$$
$$\delta = \frac{\omega_1^2 b L}{2\rho_0 c_0^3} \approx 0.3 \cdot 10^{-3}$$
$$\eta = \frac{\rho_0 l c_0}{b} \approx 3.3 \cdot 10^{-2}$$
$$q = \frac{c_0^4 l(\gamma+1)\rho_0^2}{b^2 L \omega_1^2} \approx 120\ .$$

It is obvious that these values of δ, ν and q satisfy the requirements necessary for the consistency of the approximations made for finding the solution given by (8.36), (8.38), (8.83), (8.111) and (8.115).

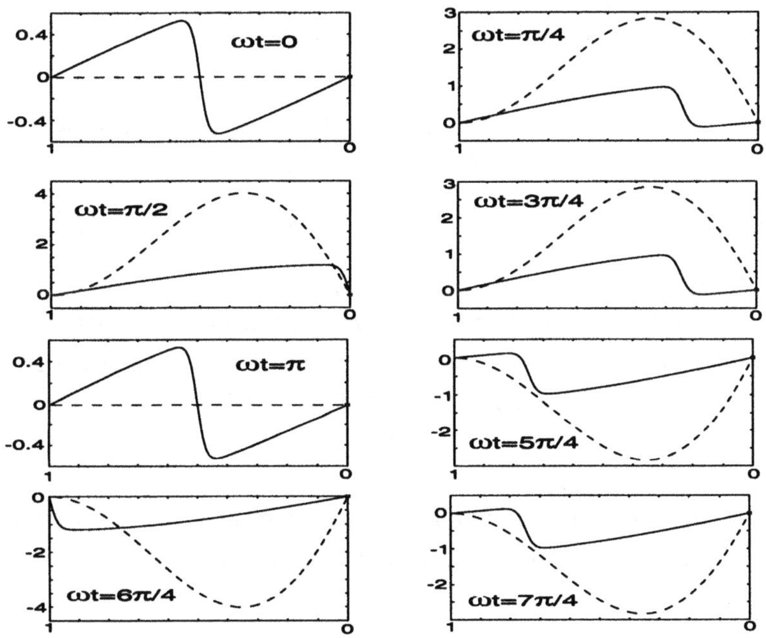

Figure 8.4: The wave amplitude along the length of a resonator for different ωt showing the wave form, where x is the distance from the moving boundary. The solid line is the nonlinear solution which is compared to the linear solution (dashed line).

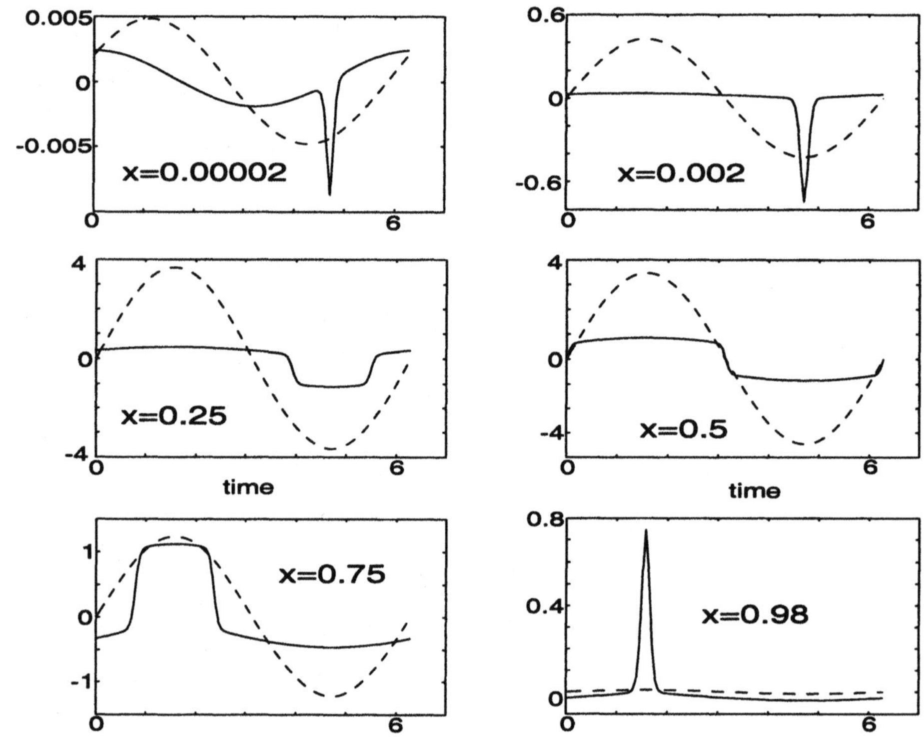

Figure 8.5: One time period of the wave at different distances x from the moving boundary in a resonator of length 1. The solid line is the nonlinear solution which is compared to the linear solution (dashed line).

Curves of this solution with the parameters above at exact resonance ($\Delta = 0$) are given in Figures 8.4 and 8.5.

Bibliography

[1] Aanonsen, S.I., Barkve, T., Naze Tjøtta, J. and Tjøtta, S., Distortion and harmonic generation in the nearfield of a finite amplitude sound beam, *J. Acoust. Soc. Am.* **75**, 749-768 (1984).

[2] Abramowitz, M. and Stegun, I.A. (eds.) *Handbook of Mathematical Functions* (National Bureau of Standards, Washington D.C., 1964).

[3] Airy, G.B., On a difficulty in the problem of sound, *Phil.Mag.* (Series 3) **34**, 401-405 (1849), or Beyer (1984), pp. 37-41.

[4] d'Alembert, J. -le- R., Investigation of a curve formed by a vibrating string (1747), translated in Lindsay (1972), pp. 119-130.

[5] Averkiou, M.A. and Cleveland, R.O., Modeling of an electrohydraulic lithotripter with the KZK equation, *J. Acoust. Soc. Am.* **106**, 102-112 (1999).

[6] Bakhvalov, N.S., Zhileikin, Ya.M., Zabolotskaya, E.A. and Khokhlov, R.V., Nonlinear propagation of a sound beam in a nondissipative medium, *Sov. Phys. Acoust.* **22**, 272-274 (1976).

[7] Bakhvalov, N.S., Zhileikin, Y.M. and Zabolotskaya, E.A., *Nonlinear Theory of Sound Beams* (American Institute of Physics, New York, 1987).

[8] Banta, E.D., Lossless propagation of one-dimensional finite-amplitude sound waves, *J. Math. Analysis and Appl.* **10**, 166-173 (1965).

[9] Bateman, H., Some recent researches on the motions of fluids, *Monthly Weather Review* **43**, 163-170 (1915).

[10] Bednarik, M. and Cervenka, M., Nonlinear waves in resonators, in *Nonlinear Acoustics at the Turn of the Millennium, ISNA 15*, eds. W. Lauterborn and Th. Kurz, (AIP Conference Proceedings, Vol. 524, New York, 2000) pp. 165-168.

[11] Bender, C.M. and Orszag, S.A., *Advanced Mathematical Methods for Scientists and Engineers* (McGraw-Hill, New York, 1978).

[12] Bennett, M.B. and Blackstock, D.T., Parametric array in air, *J. Acoust. Soc. Am.* **57**, 562-568 (1975).

[13] Berktay, H.O., Parametric amplification by the use of acoustic nonlinearities and some possible applications, *J. Sound Vib.* **2**, 462-470 (1965).

[14] Berktay. H.O. and Al-Temini, C.A., Virtual arrays for underwater reception, *J. Sound Vib.* **9**, 295-307 (1969).

[15] Berktay. H.O. and Al-Temini, C.A., Scattering of sound by sound, *J. Acoust. Soc. Am.* **50**, 181-187 (1971).

[16] Beyer, R.T., Parameter of nonlinearity in fluids, *J. Acoust. Soc. Am.* **32**, 719-721 (1959).

[17] Beyer, R.T. *Nonlinear Acoustics in Fluids* (Van Nostrand Reinhold, New York, 1984).

[18] Beyer, R.T., The parameter *B/A*, in Hamilton, M.F. and Blackstock, D.T. (eds.), *Nonlinear Acoustics* (Academic Press, San Diego, 1997), pp. 25-40.

[19] Bjørnø, L., Nonlinear acoustics, in *Acoustics and Vibration Progress, Vol. 2*, eds. R.W.B. Stephens and H.G. Leventhall (Chapman Hall, London, 1976), pp. 101-203.

[20] Bjørnø, L., Characterization of biological media by means of their nonlinearity, *Ultrasonics* **24**, 254-259 (1986).

[21] Blackstock, D.T., Propagation of plane sound waves of finite-amplitude in non-dissipative fluids, *J. Acoust. Soc. Am.* **34**, 9-30 (1962).

[22] Blackstock, D.T., Thermoviscous attenuation of plane, periodic, finite-amplitude sound waves, *J. Acoust. Soc. Am.* **36**, 534-542 (1964).

[23] Blackstock, D.T., Connection between the Fay and Fubini solutions for plane sound waves of finite amplitude, *J. Acoust. Soc. Am.* **39**, 1019-1026 (1966).

[24] Blackstock, D.T., History of nonlinear acoustics, in Hamilton, M.F. and Blackstock, D.T. (eds.), *Nonlinear Acoustics* (Academic Press, San Diego, 1997[a]), pp. 1-24.

[25] Blackstock, D.T., Hamilton, M.F. and Pierce, A.D., Progressive waves in lossless and lossy fluids, in Hamilton, M.F. and Blackstock, D.T. (eds.), *Nonlinear Acoustics* (Academic Press, San Diego, 1997), pp. 65-150.

[26] Blackstock, D.T., Audio application of the parametric array, *J. Acoust. Soc. Am.* **102**, 3106 (1997[b]).

[27] Brander, O. and Hedenfalk, J., A new formulation of the general solution to Burgers' equation, *Wave Motion* **28**, 319-332 (1998).

[28] Brysev, A.P., Krutyansky, L.M., Preobrazhensky V.L., Pylnov Yu.V., Cunningham, K.B. and Hamilton, M.F., Nonlinear propagation of phase-conjugate focused sound beams in water, in *Nonlinear Acoustics at the Turn of the Millennium,*

ISNA 15, eds. W. Lauterborn and Th. Kurz (AIP Conference Proceedings, Vol. 524, New York, 2000), pp. 183-186.

[29] Burgers, J.M., A mathematical model illustrating the theory of turbulence, in *Advances in Applied Mechanics*, Vol. 1, eds. R. von Mises and T. von Kármán (Academic Press, New York, 1948), pp. 171-199.

[30] Burgers, J.M., *The Nonlinear Diffusion Equation* (Reidel, Dordrecht, 1974).

[31] Cahill, M.D. and Baker, A.C., Oscillations in harmonics generated by the interaction of acoustic beams, *J. Acoust. Soc. Am.* **105**, 1575-1583 (1999).

[32] Campos, L.M.B.C., On waves in gases. Part 1: Acoustics of jets, turbulence and ducts, *Rev. Mod. Phys.* **58**, 117-182 (1986).

[33] Carstensen, E.L., Law, W.K., McKay, N.D. and Muir, T.G., Demonstration of nonlinear acoustical effects at biomedical frequencies and intensities, in *Ultrasound in Med. & Biol. Vol. 6* (Pergamon Press, 1980), pp. 359-368.

[34] Cary, B.B., Nonlinear losses induced in spherical waves, *J. Acoust. Soc. Am.* **42**, 88-92 (1967).

[35] Cary, B.B., Prediction of finite-amplitude waveform distortion with dissipation and spreading loss, *J. Acoust. Soc. Am.* **43**, 1364-1372 (1968).

[36] Cary, B.B., Modification of the Bessel-Fubini solution that includes attenuation, *J. Acoust. Soc. Am.* **49**, 1687-1688 (1971).

[37] Cary, B.B., An exact shock wave solution to Burgers' equation for parametric excitation of the boundary, *J. Sound Vib.* **30**, 455-464 (1973).

[38] Cary, B.B. and Fenlon, F.H., On the near and far-field radiation patterns generated by the nonlinear interaction of two separate and non-planar monochromatic sources, *J. Sound Vib.* **26**, 209-222 (1973).

[39] Cary, B.B., Asymptotic Fourier analysis of a "sawtooth like" wave for a dual frequency source excitation, *J. Sound Vib.* **42**, 235-241 (1975).

[40] Chester, W., Resonant oscillations in closed tubes, *J. Fluid Mech.* **18**, 44-66 (1964).

[41] Cobb, W.N., Finite amplitude method for determination of the acoustic nonlinearity parameter B/A, *J. Acoust. Soc. Am.* **73**, 1525-1531 (1983).

[42] Cole, J.D., On a quasi-linear parabolic equation occurring in aerodynamics, *Quart. Appl. Math.* **9**, 225-236 (1951).

[43] Cook, B.D., New procedure for computing finite-amplitude distortion, *J. Acoust. Soc. Am.* **34**, 941-946 (1962).

[44] Coppens, A.B., Beyer, R.T., Seiden, M.B., Donohue, J., Guepin, F., Hodson, R.H. and Townsend, C., Parameter of nonlinearity in fluids. II, *J. Acoust. Soc. Am.* **38**, 797-804 (1965).

[45] Coppens, A.B. and Sanders, J.V., Finite-amplitude standing waves in rigid-walled tubes, *J. Acoust. Soc. Am.* **43**, 516-529 (1968).

[46] Coppens, A.B. and Sanders, J.V., Finite-amplitude standing waves within real cavities, *J. Acoust. Soc. Am.* **58**, 1133-1140 (1975).

[47] Coppens, A.B. and Atchley, A.A., Nonlinear standing waves in cavities, in *Encyclopedia of Acoustics*, ed. M.J. Crocker (Wiley, New York, 1997), pp. 237-247.

[48] Coulouvrat, F., Solutions approchées de l'équation de Burgers par une méthode asymptotique mixte, *C. R. Acad. Sci. Paris* t. **308** (Série II), 1765-1770 (1989).

[49] Coulouvrat, F., Plane sound waves of finite amplitude for intermediate Gol'dberg numbers, *J. Acoust. Soc. Am.* **89**, 2640-2651 (1991[a]).

[50] Coulouvrat, F., An analytical approximation of strong nonlinear effects in bounded beams, *J. Acoust. Soc. Am.* **90**, 1592-1600 (1991[b]).

[51] Coulouvrat, F. Méthodes Asymptotiques en Acoustique Nonlinéaire, Thèse de doctorat de l'Université Paris 6, (1991[c]).

[52] Coulouvrat, F., On the equations of nonlinear acoustics, *J. Acoustique* **5**, 321-359 (1992).

[53] Crighton, D.G., Model equations of nonlinear acoustics, *Ann. Rev. Fluid Mech.* **11**, 11-23, (1979).

[54] Crighton, D.G. and Scott J.F., Asymptotic solutions of model equations in nonlinear acoustics, *Phil. Trans. Roy. Soc. Lond.* **A292**, 101-134 (1979).

[55] Cruikshank, D.B., Experimental investigation of finite-amplitude acoustic oscillations in a closed tube, *J. Acoust. Soc. Am.* **52**, 1024-1036 (1972).

[56] Cunningham, K.B., Hamilton, M.F., Brysev, A.P. and Krutyansky, L.M., Time-reversed sound beams of finite amplitude, *J. Acoust. Soc. Am.* **109**, 2668-2674 (2001).

[57] Darvennes, C.M., Hamilton, M.F., Naze Tjøtta, J. and Tjøtta, S., Effects of absorption on the nonlinear interaction of sound beams, *J. Acoust. Soc. Am.* **89**, 1028-1036 (1991).

[58] Derode, A., Roux, P. and Fink, M., Robust acoustic time reversal with high-order multiple scattering, *Phys. Rev. Lett.* **75**, 4206-4209 (1995).

[59] Dorme, C. and Fink M., Focusing of transmit-receive mode through inhomogeneous media: The time reversal matched filter approach, *J. Acoust. Soc. Am.* **98**, 1155-1162 (1995).

[60] Dybedal, J. Topas: Parametric end-fire array used in offshore applications, in *Advances in Nonlinear acoustics, ISNA 13*, ed. H. Hobæk (World Scientific, Singapore, 1993), pp. 264-269.

[61] Earnshaw, S., On the mathematical theory of sound, *Brit. Assn. Adv. Sci., Report of the 28th Meeting. Notices and Abstracts Sec. 34-35* (1858).

[62] Earnshaw, S., On the mathematical theory of sound, *Phil.Trans. Roy. Soc. Lond.* **150**, 133-148 (1860).

[63] Enflo, B.O., Saturation of a nonlinear cylindrical sound wave generated by a sinusoidal source, *J. Acoust. Soc. Am.* **77**, 54-60 (1985^a).

[64] Enflo, B.O., Nonlinear sound waves from a uniformly moving point source, *J. Acoust. Soc. Am.* **77**, 2054-2060 (1985^b).

[65] Enflo, B.O., The decay of the shockwave from a supersonic projectile, *AIAA Journal* **23**, 1824-1826 1826 (1985^c).

[66] Enflo, B.O., Saturation of nonlinear spherical and cylindrical sound waves, *J. Acoust. Soc. Am.* **99**, 1960-1964 (1996).

[67] Enflo, B.O., On the connection between the asymptotic waveform and the fading tail of an initial N-wave in nonlinear acoustics, *Acustica-Acta Acustica* **84**, 401-413 (1998).

[68] Enflo, B.O., Sound beams with shockwave pulses, *Acoust. Phys.* **46**, 728-733 (2000).

[69] Enflo, B.O. and Hedberg, C.M., Fourier decomposition of a plane nonlinear sound wave developing from a sinusoidal source, *Acustica-Acta Acustica* **87**, 163-169 (2001).

[70] Euler, L., De la propagation du son, *Mém. Acad. Sci. Berlin* **15**, 185-209 (1766).

[71] Fay, R. D., Plane sound waves of finite-amplitude, *J. Acoust. Soc. Am.* **3**, 222-241 (1931).

[72] Fenlon, F.H., An extension of the Fubini series for a multiple-frequency CW acoustic source of finite amplitude, *J. Acoust. Soc. Am.* **51**, 284-289 (1972).

[73] Fenlon, F.H., Derivation of the multiple frequency Bessel-Fubini series via Fourier analysis of the preshock time waveform, *J. Acoust. Soc. Am.* **53**, 1752-1754 (1973^a).

[74] Fenlon, F.H., On the derivation of a Lagrange-Banta operator for progressive finite-amplitude wave propagation in a dissipative fluid medium, *J. Acoust. Soc. Am.* **54**, 92-95 (1973^b).

[75] Fenlon, F.H., On the performance of a dual frequency parametric source via matched asymptotic solutions of Burgers' equation, *J. Acoust. Soc. Am.* **55**, 35-46 (1974).

[76] Fink, M., Time reversal acoustics, *Physics Today* **50**(3), 34-40, March 1997.

[77] Foda, M.A., Distortion and dispersion of nonlinear waves in a rectangular duct due to a bifrequency excitation, *Acustica-Acta Acustica* **82**, 411-422 (1996).

[78] Fourier, J., *The Analytical Theory of Heat* (1822). Translated by A. Freeman 1878. Reprinted by Dover, New York, 1955.

[79] Fox, F.E., and Wallace, W.A., Absorption of finite amplitude sound waves, *J. Acoust. Soc. Am.* **26**, 994-1006 (1954).

[80] Frøysa, K.-E., and Coulouvrat, F., A renormalization method for nonlinear pulsed sound beams, *J. Acoust. Soc. Am.* **99**, 3319-3328 (1996).

[81] Fubini-Ghiron, E., Anomalie nella propagazione di onde acustiche di grande ampiezza, *Alta Frequenza* **4**, 530-581 (1935). English translation: Beyer (1984), pp. 118-177.

[82] Gaitan, D.F. and Atchley, A.A., Finite amplitude standing waves in harmonic and anharmonic tubes, *J. Acoust. Soc. Am.* **93**, 2489-2495 (1993).

[83] Ginsberg, J.H., Perturbation methods, in Hamilton, M.F. and Blackstock, D.T. (eds.), *Nonlinear Acoustics*, (Academic Press, San Diego, 1997), pp. 279-308.

[84] Gradshtein, I. S. and Ryzhik I. M., *Tables of Integrals, Series and Products* (Academic Press, New York, 1965).

[85] Gurbatov, S.N., Malakhov, A.N. and Saichev, A.I., *Nonlinear Random Waves in Nondispersive Media* (Manchester University Press, Manchester, 1991).

[86] Gurbatov, S.N. and Hedberg, C.M., Nonlinear crosstransformation of amplitude-frequency modulation of quasimonochromatic signals, *Acustica - Acta Acustica* **84**, 414-424 (1998).

[87] Gurbatov, S.N., Enflo, B.O. and Pasmanik, G.V., The decay of pulses with complex structure according to Burgers' equation, *Acustica-Acta Acustica* **85**, 181-196 (1999).

[88] Gurbatov, S.N., Enflo, B.O. and Pasmanik, G.V., The decay of plane wave pulses with complex structure in a nonlinear dissipative medium, *Acustica-Acta Acustica* **87**, 16-28 (2001).

[89] Gusev, V.E., Buildup of forced oscillations in acoustic oscillators, *Sov. Phys. Acoust.* **30**, 121-125 (1984).

[90] Hamilton, M.F., Naze Tjøtta, J. and Tjøtta, S, Nonlinear effects in the farfield of a directive sound source, *J. Acoust. Soc. Am.*.**78**, 202-216 (1985).

[91] Hamilton, M.F. and Blackstock, D.T. (eds.), *Nonlinear Acoustics*, (Academic Press, San Diego, 1997).

[92] Hamilton, M.F. and Morfey, C.L., Model equations, in Hamilton, M.F. and Blackstock, D.T. (eds.), *Nonlinear Acoustics* (Academic Press, San Diego, 1997), pp. 41-64.

[93] Hamilton, M.F., Sound beams, in Hamilton, M.F. and Blackstock, D.T. (eds.), *Nonlinear Acoustics* (Academic Press, San Diego 1997), pp. 233-262.

[94] Hamilton, M.F., Khokhlova, V.A. and Rudenko, O.V., Analytical method for describing the paraxial region of finite amplitude sound beams, *J. Acoust. Soc. Am.***101**, 1298-1308 (1997).

[95] Hammerton, P.W. and Crighton, D.G., Old-age behaviour of cylindrical and spherical waves: numerical and asymptotic results, *Proc. Roy. Soc. Lond.* **A422**, 387-405 (1989).

[96] Hart, T.S. and Hamilton, M.F., Nonlinear effects in focused sound beams, *J. Acoust. Soc. Am.* **84**, 1488-1496 (1988).

[97] Hedberg, C.M., Nonlinear propagation through a fluid of waves originating from a biharmonic sound source, *J. Acoust. Soc. Am.* **96**, 1821-1828 (1994[a]).

[98] Hedberg, C.M., Theoretical Studies of Nonlinear Propagation of Modulated Harmonic Sound Waves, Doctoral thesis, *TRITA-MEK 1994-10*, Kungl. Tekniska Högskolan, Stockholm (1994[b]).

[99] Hedberg, C.M., Influence of the phase between original frequencies on the nonlinear generation of new harmonics, *J. Acoust. Soc. Am.* **99**, 3329-3333 (1996).

[100] Hedberg, C.M., Solving the inverse problem in nonlinear acoustics by backpropagation of the received signal, *J. Acoust. Soc. Am.* **101**, 3090 (1997).

[101] Hedberg, C.M., Parameter sensitivity in nonlinear and dissipative time-reversed acoustics, in *Proceedings of International Congress of Acoustics/ASA, Seattle 20-26 June 1998*, pp. 543-544 (1998).

[102] Hedberg, C.M., Multi-frequency plane, nonlinear and dissipative waves at arbitrary distances, *J. Acoust. Soc. Am.* **106**, 3150-3155 (1999).

[103] Helmholtz, H., 1. Theoretische Akustik, *Fortschritte der Physik im Jahre 1848* **4** 101-118 (1852).

[104] Hobæk, H., Parametric acoustic transmitting arrays and a survey of theories and experiments, *Scientific Technical Report No 99*, University of Bergen, Norway, (1977).

[105] Hochstadt, H., *The Functions of Mathematical Physics*, (Wiley, New York, 1971).

[106] Hopf, E., The partial differential equation $u_t + uu_x = \mu u_{xx}$, *Comm. Pure Appl. Math.* **3**, 201-230 (1950).

[107] Hugoniot, H., Mémoire sur la propagation du mouvement dans les corps et spécialement dans les gaz parfaits, *J. l'école polytech.* (Paris) **57**, 3-97 (1887), and *J. l'école polytech.* (Paris) **58**, 1-125 (1889).

[108] Hunt, F.V., Notes on the exact equations governing the propagation of sound in fluids, *J. Acoust. Soc. Am.* **27**, 1019 (1955)

[109] Ilgamov, M.A., Zaripov, R.G., Galiullin, R.G. and Repin, V.B., Nonlinear oscillations of a gas in a tube, *Appl. Mech. Rev.* **45**, 137-154 (1996).

[110] Ilinskii, Y., Lipkens, B., Lucas, T.S., Van Doren, T.W. and Zabolotskaya, E.A., Nonlinear standing waves in an acoustical resonator, *J. Acoust. Soc. Am.* **104**, 2664-2674 (1998).

[111] Ingard, U. and Pridmore-Brown, D.C., Scattering of sound by sound, *J. Acoust. Soc. Am.* **28**, 367-369 (1956).

[112] Jackson, D.R. and Dowling, D.R., Phase conjugation in underwater acoustics, *J. Acoust. Soc. Am.* **89**, 171-181 (1991).

[113] Jimenez, J., Nonlinear gas oscillations in pipes. Part 1. Theory, *J. Fluid. Mech.* **59**, 23-46 (1973).

[114] Kamakura, T., Ikegaya, K. and Chou, I-M., Nonlinear interactions of finite amplitude and weak subharmonic plane waves - Phase dependent parametric amplification, *J. Acoust. Soc. Jpn* (E) **6**, 155-160 (1985).

[115] Karabutov, A.A., Lapshin E.A.and Rudenko, O.V., Interaction between light waves and sound under acoustic nonlinear conditions, *Sov. Phys. JETP*, **44**, 58-63 (1980).

[116] Karabutov, A.A. and Rudenko, O.V., Nonlinear plane waves excited by volume sources in a medium moving with transonic velocity, *Sov. Phys. Acoust.*, **25**, 306-309 (1980).

[117] v. Kármán, T., Supersonic aerodynamics: principles and applications, *J. Aeronaut. Sci.* **14**, 373-409 (1947).

[118] Keller, J., Resonant oscillations in closed tubes: the solution of Chester's equation, *J. Fluid Mech.* **77**, 279-304 (1976).

[119] Kirchhoff, G., On the influence of heat conduction in a gas on sound propagation, *Ann. Phys. Chem. (5)* **134**, 177-193 (1868).

[120] Kuperman, W.A., Hodgkiss, W.S., Song, H.C., Akal, T., Ferla, C. and Jackson, D.R., Phase-conjugation in the ocean: experimental demonstration of an acoustic time reversal mirror, *J. Acoust. Soc. Am.* **103**, 25-40 (1998).

[121] Kuznetsov, V.P., Equations of nonlinear acoustics, *Sov. Phys. Acoust.* **16**, 467-470 (1971).

[122] Lagrange, J.L., Sec. 42 in Nouvelles recherces sur la nature et la propagation du son, *Miscellanea Taurinensis* II, 11-172 (1760-61).

[123] Lardner, R.W., Acoustic saturation and the conversion efficiency of the parametric array, *J. Sound Vib.* **82**, 473-487 (1982).

[124] Lawrenson, C.C., Lipkens, B., Lucas, T.S., Perkins, D.K. and Van Doren, T.W., Measurement of macrosonic standing waves in oscillating closed cavities, *J. Acoust. Soc. Am.* **104**, 623-636 (1998).

[125] Lebedev, N.N., *Special Functions and their Applications* (Prentice-Hall, Englewood Cliffs, 1965).

[126] Lee, Y.-S. and Hamilton, M.F., Time-domain modeling of pulsed finite amplitude sound beams, *J. Acoust. Soc. Am.* **97**, 906-917 (1995).

[127] Lesser, M.B. and Crighton, D.G., Physical acoustics and the method of matched asymptotic expansions, in *Physical Acoustics*, eds. W.P. Mason and R.N. Thurston (Academic Press, New York, 1975), pp. 69-149.

[128] Lighthill, M.J., Viscocity effects in sound waves of finite amplitude, in *Surveys in Mechanics*, eds. G.K. Batchelor and R.M. Davies (Cambridge University Press, 1956), pp. 250-351.

[129] Lindsay, R.B., *Acoustics: Historical and Philosophical Development* (Dowden, Hutchinson & Ross, Stroudsburg, Pa., 1972).

[130] Makov, Yu.N., Universal automodeling solution to the Khokhlov-Zabolotskaya equation for waves with shock fronts, *Acoust. Phys.* **43**, 722-727 (1997).

[131] McLachlan, N.W., *Theory and Application of Mathieu Functions* (Oxford University Press, 1947).

[132] Mendousse, J.S., Nonlinear dissipative distortion of progressive sound waves at moderate amplitudes, *J. Acoust. Soc. Am.* **25**, 51-54 (1953).

[133] Moffett, M.B. and Mellen, R.H., Model for parametric acoustic sources, *J. Acoust. Soc. Am.* **61**, 325-337 (1977).

[134] Naugolnykh, K.A., Propagation of spherical sound waves of finite-amplitude in a viscous heat-conducting medium, *Sov. Phys. Acoust.* **5**, 79-84 (1959).

[135] Naugolnykh, K.A., Soluyan, S.I. and Khokhlov, R.V., Cylindrical waves of finite-amplitude in a dissipative medium, *Vestn. Moscow State Univ., Fiz. Astron.* **4**, 65-71 (1962).

[136] Naugolnykh, K.A., Soluyan, S.I. and Khokhlov, R.V., Speherical waves of finite-amplitude in a viscous thermally conducting medium, *Sov. Phys. Acoust.* **9**, 42-46 (1963[a]).

[137] Naugolnykh, K.A., Soluyan, S.I. and Khokhlov, R.V., Nonlinear interaction of sound waves in an absorbing medium, *Sov. Phys. Acoust.* **9**, 155-159 (1963[b]).

[138] Naugolnykh, K.A., Transition of a shock wave into a sound wave, *Sov. Phys. Acoust.* **18**, 475-477 (1973).

[139] Naugolnykh, K.A. and Ostrovsky, L., *Nonlinear Wave Processes in Acoustics* (Cambridge University Press, 1998).

[140] Navier, L.M.H., Mémoire sur les lois du mouvement des fluids, *Mém. Acad. Sci.* **6**, 389-416 (1823).

[141] Nayfeh, A. H., *Perturbation Methods* (Wiley-Interscience, New York, 1973).

[142] Naze Tjøtta, J. and Tjøtta, S., Nonlinear equations of acoustics, with application to parametric acoustic arrays, *J. Acoust. Soc. Am.* **69**, 1644-1652 (1981).

[143] Naze Tjøtta, J. and Tjøtta, S., Nonlinear equations of acoustics, in *Frontiers of Nonlinear Acoustics, ISNA 12*, eds. M.F. Hamilton and D.T. Blackstock (Elsevier, London, 1990), pp. 80-97.

[144] Naze Tjøtta, J., Tjøtta, S. and Vefring, E.H., Effects of focusing on the nonlinear interaction between two collinear finite amplitude sound beams, *J. Acoust. Soc. Am.* **89**, 1017-1027 (1991).

[145] Naze Tjøtta, J., TenCate, J.A. and Tjøtta, S., Effects of boundary conditions on the nonlinear interaction of sound beams, *J. Acoust. Soc. Am.* **89**, 1037-1049 (1991).

[146] Ni, A.L., Non-linear resonant oscillations of a gas in a tube under the action of a periodically varying pressure, *Prikl. Matem. Mekhan.* **47**, 607-618 (1983) translated in *PMM USSR* **47**, 498-506 (1983).

[147] Novikov, B.K. and Rudenko, O.V., Degenerate amplification of sound, *Sov. Phys. Acoust.* **22**, 258-259 (1976).

[148] Novikov, B.K., Exact solutions of the Burgers equation, *Sov. Phys. Acoust.* **24**, 326-328 (1978).

[149] Novikov, B.K., Rudenko, O.V. and Timoshenko, V.I., *Nonlinear underwater acoustics*, (American Institute of Physics, New York, 1987)

[150] Nyberg, Ch., Spectral analysis of a two frequency driven resonance in a closed tube, *Acoust. Phys.* **45**, 94-104 (1999).

[151] Parker, A., On the periodic solution of the Burgers equation: a unified approach, *Proc. Roy. Soc. Lond.* **A438**, 113-132 (1992).

[152] Parker, D.F., The decay of sawtooth solutions to the Burgers equation, *Proc. Roy. Soc. Lond.* **A369**, 409-424 (1980).

[153] Pierce, A.D., *Acoustics* (McGraw-Hill, New York, 1981).

[154] Pierce, A.D., Nonlinear acoustic research topics stimulated by the sonic boom problem, in *Advances in Nonlinear acoustics, ISNA 13*, ed. H. Hobæk (World Scientific, Singapore, 1993), pp. 7-20.

[155] Poisson, S.D., Mémoire sur la théorie du son, *J. l'école polytech.* (Paris) **7**, 319-392 (1808).

[156] Rankine, W.J.M., On the thermodynamic theory of waves of finite longitudinal disturbance, *Phil. Trans. Roy. Soc.* **160**, 277-288 (1870), or Beyer (1984), pp. 65-76.

[157] Rayleigh, Lord, Aerial plane waves of finite amplitude, *Proc. Roy. Soc. Lond.* **A84**, 247-284 (1910).

[158] Riemann, B., Ueber die Fortplanzung ebener Luftwellen von endlicher Schwingungsweite, *Abhandl. Ges. Wiss. Göttingen, Math.-Physik.* **8**, 43-65 (1860).

[159] Rossing, T.D., *The Science of Sound*, 2nd ed., (Addison-Wesley, 1990).

[160] Rott, N., The description of simple waves by particle displacement, *Z. angew. Math. Phys.* **29**, 178-189 (1978).

[161] Rott, N., Nonlinear acoustics, in *Theoretical and Applied Mechanics, 15th International Congress of Theoretical and Applied Mechanics - ICTAM - Toronto, August 17-23, 1980*, eds. F.P.J. Rimrott and B. Tabarrok (North-Holland, 1980), pp. 163-173.

[162] Rudenko, O.V., Soluyan, S.I. and Khokhlov, R.V., Problems in the theory of nonlinear acoustics, *Sov. Phys. Acoust.* **20**, 271-275 (1974).

[163] Rudenko, O.V., Feasability of generation of high-power hypersound with the aid of laser radiation, *JETP Lett.* **20**, 203-204 (1974).

[164] Rudenko, O.V., Soluyan, S.I. and Khokhlov, R.V., Nonlinear theory of paraxial sound beams, *Sov. Phys. Dokl.* **20**, 836-837 (1976).

[165] Rudenko, O.V. and Soluyan, S.I., *Theoretical Foundations of Nonlinear Acoustics* (Plenum, New York, 1977).

[166] Rudenko, O.V. and Khokhlova, V., Kinetics of one-dimensional sawtooth waves, *Sov. Phys. Acoust.* **37**, 90-93 (1991).

[167] Rudenko, O.V., Nonlinear sawtooth-shaped waves, *Physics-Uspekhi* **38**, 965-989 (1995).

[168] Rudenko, O.V., Nonlinear oscillations of linearly deformed medium in a closed resonator excited by finite displacements of its boundary, *Acoust. Phys.* **45**, 351-356 (1999).

[169] Rudenko, O.V. and Shanin, A.V., Nonlinear phenomena accompanying the development of oscillations excited in a layer of a linear dissipative medium by finite displacements of its boundary, *Acoust. Phys.* **46**, 334-341 (2000).

[170] Rudenko, O.V. and Enflo, B.O., Nonlinear N-wave propagation through a one-dimensional phase screen, *Acustica-Acta Acustica* **86**, 229-238 (2000).

[171] Rudenko, O.V., Hedberg, C.M. and Enflo, B.O., Nonlinear standing waves in a layer excited by the periodic motion of its boundary, *Acoust. Phys.* **47**, 525-533 (2001).

[172] Sachdev, P.L. and Seebass, A.R., Propagation of spherical and cylindrical *N*-waves, *J. Fluid Mech.* **58**, 197-205 (1973).

[173] Sachdev, P.L., Tikekar, V.G. and Nair, K.R.C., Evolution and decay of spherical and cylindrical N-waves, *J. Fluid Mech.* **172**, 347-371 (1986).

[174] Sachdev, P.L., *Nonlinear Diffusive Waves*, (Cambridge University Press, 1987).

[175] Sachdev, P.L. and Nair, K.R.C., Evolution and decay of cylindrical and spherical nonlinear acoustic waves generated by a sinusoidal source, *J. Fluid Mech.* **204**, 389-404 (1989).

[176] Sachdev, P.L., Joseph, K.T. and Nair, K.R.C., Exact N-wave solutions for the non-planar Burgers equation, *Proc. Roy. Soc. Lond.* **A445**, 501-517 (1994).

[177] Scott, J.F., Uniform asymptotics for spherical and cylindrical nonlinear acoustic waves generated by a sinusoidal source, *Proc. Roy. Soc. Lond.* **A375**, 211-230 (1981).

[178] Shooter, J.A., Muir, T.G. and Blackstock, D.T., Acoustic saturation of spherical waves in water, *J. Acoust. Soc. Am.* **55**, 54-62 (1974).

[179] Sionoid, P., Nonlinear acoustic beams and the Zabolotskaya-Khokhlov equation. A report for the Senate of the National University of Ireland, DAMTP, University of Cambridge, (April 1992).

[180] Sionoid, P., The generalized Burgers and Zabolotskaya-Khokhlov equations: transformations, exact solutions and qualitative properties, in *Advances in Nonlinear Acoustics, ISNA 13*, ed. H. Hobæk (World Scientific, Singapore, 1993), pp. 63-67.

[181] Söderholm, L.H., A higher order acoustic equation for the slightly viscous case, *Acustica-Acta Acustica* **87**, 29-33 (2001).

[182] Soluyan, S. I. and Khokhlov R. V., Propagation of acoustic waves of finite amplitude in a dissipative medium, *Vestn. Mosk. Univ. Fiz. Astron.* **3**, 52-61 (1961).

[183] Stokes, G.G., On the theory of internal friction of fluids in motion, etc., *Cambridge Trans.* **8**, 287-305 (1845).

[184] Stokes, G.G., On a difficulty in the theory of sound, *Phil.Mag.* (Series 3) **33**, 349-356 (1848), or Beyer (1984), pp. 29-36.

[185] Taylor, G.I., The conditions necessary for discontinuous motion in gases, *Proc. Roy. Soc. Lond.* **A84**, 371-377 (1910).

[186] Taylor, G.I., *Scientific Papers* (Cambridge University Press, 1963).

[187] Temkin, S., Nonlinear gas oscillations in a resonant tube, *Phys. Fluids* **11**, 960-963 (1968).

[188] Temkin, S., Propagating and standing sawtooth waves, *J. Acoust. Soc. Am.* **45**, 224-227 (1969^a).

[189] Temkin, S., Attenuation of guided, weak sawtooth waves, *J. Acoust. Soc. Am.* **46**, 267-271 (1969^b).

[190] Temkin, S. and Maxham, D., Nonlinear lengthening of a triangular acoustic pulse, *Phys. Fluids* **28**, 3013-3017 (1985).

[191] Trivett, D.H. and Van Buren, A.L., Propagation of plane, cylindrical and spherical finite amplitude waves, *J. Acoust. Soc. Am.* **69**, 943-949 (1981).

[192] Van Buren, A.L. and Breazeale, M.A., Reflection of finite-amplitude ultrasonic waves. II. Propagation, *J. Acoust. Soc. Am.* **44**, 1021-1027 (1968).

[193] Van Dyke, M.D. *Perturbation Methods in Fluid Mechanics* (Parabolic Press, Stanford, 1975).

[194] Vanhille, C. and Campos-Pozuelo, C., Numerical model for nonlinear standing waves and weak shocks in thermoviscous fluids, *J. Acoust. Soc. Am.* **109**, 2660-2667 (2001).

[195] Westervelt, P.J., Scattering of sound by sound, *J. Acoust. Soc. Am.* **29**, 199-203 (1957).

[196] Westervelt, P.J., Parametric acoustic arrray, *J. Acoust. Soc. Am.* **35**, 535-537 (1963).

[197] Westervelt, P.J., The status and future of nonlinear acoustics, *J. Acoust. Soc. Am.* **57**, 1352-1356 (1975).

[198] Whitham, G.B., The behaviour of supersonic flow past a body of revolution, far from the axis, *Proc. Roy. Soc. Lond.* **A203**, 89-109 (1950).

[199] Whitham, G.B., The flow pattern of a supersonic projectile, *Commun. Pure. Appl. Math.* **5**, 301-348 (1952).

[200] Whitham, G.B., *Linear and Nonlinear Waves* (Wiley, New York, 1974).

[201] Whittaker, E.T and Watson, G.N., *A Course of Modern Analysis*, 4th ed., (Cambridge University Press, 1950).

[202] Van Wijngaarden, L., Nonlinear acoustics, in *Symposium on Applied Mathematics, dedicated to the late Prof. Dr. R. Timman*, (Delft University Press, 1978), pp. 51-68.

[203] Yonegama, M. and Fujimoto, J., The audio spotlight: an application of nonlinear interaction of sound waves to a new type of loudspeaker design, *J. Acoust. Soc. Am.* **73**, 1532-1536 (1983).

[204] Ystad, B. and Berntsen, J., Numerical studies of the KZK equation for focusing sources, *Acta Acustica* **3**, 323-330 (1995).

[205] Zabolotskaya, E. A. and Khokhlov, R. V., Quasi-plane waves in the nonlinear acoustics of confined beams, *Sov. Phys. Acoust.* **15**. 35-40 (1969).

[206] Zarembo, L.K. and Krasil'nikov, V.A., Radiating parametric array, *Sov. Phys. Uspekhi* **22**, 656-661 (1979).

[207] Zverev, V. and Kalachev, A.I., Sound radiation from the region of interaction of two sound beams, *Sov. Phys. Acoust.* **15**, 322-327 (1970).

Name index

Aanonsen, S.I. 200
Abramowitz, M. 233, 239,
Airy, G.B. 2
d'Alembert, J.-le-R. 1
Akal, T. 72 Al-Temini, C.A. 87
Atchley, A.A. 219, 220
Averkiou, M.A. 199, 200

Baker, A.C. 200
Bakhvalov, N.S. 9, 200, 205
Banta, E.D. 62, 153
Barkve, T. 200
Bateman, H. 3, 5
Bednarik, M. 221
Bender, C.M. 50
Bennett, M.B. 87
Berktay, H.O. 7, 87,
Berntsen, J. 212
Beyer, R.T. 4, 23, 24
Bjørnø, L. 11, 13, 24
Blackstock, D.T. 4, 6, 7, 62, 65, 87, 107, 112, 125, 133, 134, 187, 192, 196
Brander, O. 94
Breazeale, M.A. 24
Burgers, J.M. 4, 5, 26

Cahill, M.D. 200

Campos, L.M.B.C. 21
Campos-Pozuelo, C. 220
Carstensen, E.L. 199
Cary, B.B. 62, 78, 87, 147, 152, 153
Chester, W. 9, 219, 224, 232, 238
Chou, I-M. 90
Cleveland, R.O. 199, 200
Cobb, W.N. 24
Cole, J.D. 5, 6, 47, 106, 108
Cook, B.D. 133
Coppens, A.B. 24, 220,
Coulouvrat, F. 11, 112, 122, 147, 200
Crighton, D.G. 4, 7, 8, 9, 96, 150, 153, 166, 197, 215
Cruikshank, D.B. 220

Darvennes, C .M. 199
Derode, A. 72
Donohue, J. 24
Dorme, C. 72
Dowling, D.R. 73
Dybedal, J. 87

Earnshaw, S. 2
Enflo, B.O. 6, 7, 8, 9, 59, 72, 94, 100, 126, 128, 134, 154, 166, 173, 186, 189, 191, 192, 196, 197, 205, 221, 230
Euler, L. 1, 2, 15

Fay, R.D. 3, 5, 108
Fenlon, F. 7, 78, 82, 87, 142, 147, 153
Ferla, C. 72
Fink, M. 72
Foda, M.A. 199
Fourier, J. 18
Fox, F.E. 24
Frøysa, K.-E. 200
Fubini-Ghiron, E. 3, 5, 62, 76

Fujimoto, J. 87

Gaitan, D.F. 219
Galiullin, R.G. 219
Ginsberg, J.H. 200
Gradshteyn, I.S. 106, 107
Guepin, F. 24
Gurbatov, S.N. 44, 59, 72, 91, 94, 145
Gusev, V.E. 220

Hamilton, M.F. 4. 9, 73, 199, 200, 201
Hammerton, P.W. 8, 9, 166, 197
Hart, T.S. 200
Hedberg, C.M. 6, 7, 44, 73, 88, 91, 119, 126, 128, 134, 140, 145, 221, 230
Hedenfalk, J. 94
Helmholtz, H. 3
Hobæk, H. 87
Hochstadt, H. 78,
Hodgkiss, W.S. 72
Hopf, E. 5, 47
Hodson, R.H. 24
Hugoniot, H. 3, 38
Hunt, F.V. 13

Ikegaya, K. 90
Ilgamov, M.A. 219
Ilinskii, Yu.A. 220
Ingard, U. 87

Jackson, D.R. 72
Jimenez, J. 220
Joseph, K.T. 153

Kalachev, A.I. 87
Kamakura, T. 90

Karabutov, A.A. 231
v. Kármán, T. 170
Keller, J. 220
Khokhlov, R.V. 4, 5, 6, 9, 28, 29, 84, 114, 117, 149, 152, 199, 200
Khokhlova, V. 65, 200, 201
Kirchhoff, G. 18
Krasilnikov, V.A. 87
Kuperman, W.A. 72
Kuznetsov, V.P. 4, 5, 9, 23, 29, 199, 221

Lagrange, J.L. 2
Lapshin, E.A. 231
Lardner, R.W. 7, 137, 147
Law, W.K. 199
Lawrenson, C.C. 220
Lebedev, N.N. 104, 161, 164
Lee, Y.-S. 200
Lesser, M.B. 96
Lighthill, M.J. 5, 9
Lindsay, R.B. 4
Lipkins, B. 220
Lucas, T.S. 220

Makov, Yu.N. 200
Malakhov, A.N. 72
Maxham, D. 56
McKay, N.D. 199
McLachlan, N.W. 239
Mellen, R.H. 87
Mendousse, J.S. 5, 106, 137,
Moffett, M.B. 87
Morfey, C.L. 4
Muir, T.G. 7, 187, 192, 196, 199

Nair, K.R.C. 7, 8, 9, 153, 166, 197
Naugolnykh, K. 28, 84, 87, 149, 152

Navier, L.M.H. 15
Nayfeh, A.H. 96
Naze Tjøtta, J. 11, 87, 199, 200
Ni, A.L. 220
Novikov, B.K. 9, 43, 87, 89, 122, 199, 203, 204
Nyberg, C. 220

Orszag, C.A. 50

Parker, A. 122
Parker, D.F. 122
Pasmanik, G.V. 59, 72, 94
Perkins, D.K. 220
Pierce, A.D. 4, 18, 22, 33, 38, 170, 173, 180, 186
Poisson, S.D. 2
Pridmore-Brown, D.C. 87
Pythagoras 1

Rankine, W.J.M. 3
Rayleigh, Lord 3
Repin, V.B. 219
Riemann, B. 2, 31
Rossing, Th. 3
Rott, N. 42
Roux, P. 72
Rudenko, O.V. 4, 9, 11, 60, 65, 87, 89, 89, 114, 176, 186, 199, 200, 201, 203, 204, 205, 221, 230, 231
Ryzhik, I.M. 106, 107

Sachdev, P.L. 7, 8, 149, 153, 166, 197
Saichev, A.I. 72
Sanders, J.V. 220
Scott, J.F. 7, 8, 96, 150, 153, 166, 191, 196, 215
Seebass, A.R. 166
Seiden, M.B. 24
Shanin, A.V. 221

Shooter, J.A, 7, 187, 192, 196
Sionoid, P. 9, 201, 205, 205
Soluyan, S.I. 4, 6, 11, 28, 60, 84, 114, 117, 149, 152, 176
Song, H.C. 72
Stegun, I.A. 233, 239,
Stokes, G.C. 3, 15
Söderholm, L.H. 21, 23

Tartini, G. 3
Taylor, G.I. 3, 8, 98
Temkin, S. 56, 65, 220
TenCate, J.A. 199
Tikekar, V.G. 7, 8, 166
Timoshenko, V.I. 9, 87, 199, 203, 204
Tjøtta, S. 11, 87, 199, 200
Townsend, C. 24
Trivett, D.H. 153

Van Buren, A.L. 24, 153
Van Doren, T.W. 220
Van Dyke, M.D. 96
Van Wijngaarden, L. 220
Vefring, E.H. 199

Wallace, W.A. 24
Vanhille, C. 220
Watson, G.N. 108, 120
Westervelt, P.J. 4, 7, 87,
Whitham, G.B. 8, 32, 56, 167, 173, 186
Whittaker, E.T. 108, 120

Yonegama, M. 87
Ystad, B. 212

Zabolotskaya, E.A. 5, 9, 29, 199, 200, 205, 220
Zarembo, L.K. 87

Zaripov, R.G. 219
Zhileikin, Ya. M. 9, 200, 205
Zverev, V.A. 87

Subject index

Absorption distance 203
Acoustical Reynolds number 26
Area differences 43
Asymptotic matching 124, 244-247

B/A, the nonlinear parameter 24
Bessel functions, integral representation of 61, 78
Bessel functions, modified, integral representation of 69, 105
Bessel-Fubini formula 62
Bifrequency wave 76-84, 141-147
Bifurcation 234
Burgers' equation 3, 26, 27, 46, 94, 100, 211
Burgers' equation, generalized for cylindrical waves 5, 28, 150, 174, 194, 205, 211
Burgers' equation, generalized for spherical waves 5, 28, 150, 174, 188, 211
Burgers' equation, inhomogeneous 231

Characteristic curves 32, 34
Cole-Hopf transformation 47
Combination tones 3
Continuity equation 14

Diffraction length 203
Discontinuity length 201
Discrete integration 129
Dissipation parameter 138, 222

Elliptic integral 233
Energy, conservation of 15
Enthalpy 18
Entropy 17
Equal areas, rule of 38-42
Eulerian coordinates 12
Euler's fundamental hydrodynamical equations 15

Fay solution 3, 108
Fay solution, approximate 188, 194
Fay solution, improved 112
Fubini solution 3, 62
Fubini solution, corrected 125

Green function 170

Heat conduction equation 46
Heat conduction number 17
Hermite function 104, 160
Hermite polynomial 104

Ideal fluid equation of state 18
Irrotational flow 20

Khokhlov-Soluyan solution 117
Kuznetsov's equation 4, 23, 226
KZK (Khokhlov-Zabolotskaya-Kuznetsov) equation 5, 29, 201

Lagrangian coordinates 12

Mach cone 172
Mach number 170, 224
Mass, conservation of 13
Matched asymptotic expansions 96, 222-225
Mathieu equation 238

Maxwell relation 19
Momentum, conservation of 14
Multifrequency waves 84-92, 137-147

Navier-Stokes equations 14
Nonlinearity parameter of fluid 23
N-wave, cylindrical 8, 153-166
N-wave generation 178
N-wave, planar 8, 56-59, 93-104, 143
N-wave, spherical 8

Old-age constant 165
Old-age region 100, 154
Old-age solution 98, 165

Parabolic equation 204
Parametric array 87

Q-factor 237

Rankine-Hugoniot relations 3, 38
Resonance 219, 224
Resonator, nonlinear velocity field in 247-250
Riemann invariants 31
Riemann's equation 5, 32
Riemann wave 32

Saddle-point (steepest descent) method 50, 66, 101, 158
Saturation 6, 113
Sawtooth wave 64, 91
Shock distance 201
Shock preservation 214-216
Shock wave 36
Shock wave, from a supersonic projectile 166-186
Shock wave, in sound beam 210-214
Sinusoidal waves, cylindrical 193-197

Sinusoidal waves, lossless 59
Sinusoidal waves, planar 105-136
Sinusoidal waves, spherical 187-193
Sonic boom 186
Sound beam, circular 208-210
Sound beam, infinite in a transverse direction 205-208
Sound speed 32
Standing waves, linear 222-226
Standing waves, nonlinear 226-250
Strong nonlinearity 199, 200
Substantial time derivative 13

Tartini tones 3
Taylor shock 98
Theta function 108
Time reversal 72
Triangular pulse 53

Viscosity, bulk 15
Viscosity, shear 15

Weak nonlinearity 199, 200
Weak shock theory 8
Whitham's F-function 172

GPSR Compliance

The European Union's (EU) General Product Safety Regulation (GPSR) is a set of rules that requires consumer products to be safe and our obligations to ensure this.

If you have any concerns about our products, you can contact us on ProductSafety@springernature.com

In case Publisher is established outside the EU, the EU authorized representative is:

Springer Nature Customer Service Center GmbH
Europaplatz 3
69115 Heidelberg, Germany

Batch number: 10584469

Printed by Printforce, the Netherlands